MW00885548

STERLING
Test Prep

SAT

CHEMISTRY

Practice Questions

5th edition

www.Sterling-Prep.com

Copyright © 2015 Sterling Test Prep

We strive to provide the highest quality preparation materials.
Be the first to report an error, typo or inaccuracy in the content of this publication
to info@sterling-prep.com to receive a $10 reward for content error or
$5 reward for a typo or grammatical mistake.

All rights reserved. The content of this publication, including the text and graphic images, or any part thereof, may not be reproduced or distributed in any manner whatsoever without the written consent from the publisher.

5 4 3 2 1

ISBN-13: 978-1-5001586-8-2

Sterling Test Prep products are available at special quantity discounts for sales, promotions, premed counseling offices and other educational purposes.

For more information contact our Sales Department at

Sterling Test Prep
6 Liberty Square #305
Boston, MA 02109

info@sterling-prep.com

© 2015 Sterling Test Prep

Published by Sterling Test Prep

Printed in the U.S.A.

Copyright © 2015 Sterling Test Prep

Congratulations on choosing this book as part of your SAT Chemistry preparation!

Scoring well on the SAT Chemistry is important for admission into college. To achieve a high score, you need to develop skills to properly apply the knowledge you have and quickly choose the correct answer. You must solve numerous practice questions that represent the style and content of the SAT questions. Understanding key science concepts is more valuable than memorizing terms.

This book provides 1,733 chemistry practice questions that test your knowledge of all SAT Chemistry topics. At the back of the book, you will find answer keys which you should use to verify your answers. If you find that you pick too many wrong choices in any particular section, you should spend some time reviewing that topic and its key concepts.

All the questions are prepared by our science editors that possess extensive credentials, are educated in top colleges and universities. Our editors are experts on teaching sciences, preparing students for standardized science tests and have coached thousands of undergraduate and graduate school applicants on admission strategies.

We wish you great success in your future academic achievements and look forward to being an important part of your successful preparation for SAT Chemistry!

Sterling Test Prep Team

150513gdx

Your feedback is important to us because we strive to provide the highest quality test prep materials. If you are satisfied with the content of this book, share your opinion with other readers by publishing your review on Amazon.com

If you have any questions or comments about the material, email us and we will do our best to resolve any issues to your satisfaction.

Visit www.Sterling-Prep.com for SAT online practice tests

Our advanced online testing platform allows you to practice these and other SAT questions on the computer and generate Diagnostic Reports for each test.

By using our online SAT tests and Diagnostic Reports, you will be able to:

- Assess your knowledge of different topics tested on your SAT subject test
- Identify your areas of strength and weakness
- Learn important scientific topics and concepts
- Improve your test taking skills by solving numerous practice questions

To access the online tests at a special pricing go to page 232 for web address

Copyright © 2015 Sterling Test Prep

Table of Contents

Copyright © 2015 Sterling Test Prep

SAT CHEMISTRY
PRACTICE QUESTIONS

Copyright © 2015 Sterling Test Prep

Chapter 1. ELECTRONIC AND ATOMIC STRUCTURE OF MATTER; PERIODIC TABLE

1. Which choice below represents an element?

    A. glucose   **B.** sodium chloride   **C.** methanol   **D.** hydrogen   **E.** brass

2. Which of the following is a general characteristic of a metal?

    A. conductor of heat   **C.** solid state

    B. high density   **D.** malleable   **E.** all of the above

3. How many electrons can occupy the *f* subshell?

    A. 2   **B.** 6   **C.** 10   **D.** 14   **E.** 18

4. How many valence electrons are in the Lewis dot structure of C_2H_6?

    A. 10   **B.** 14   **C.** 20   **D.** 28   **E.** 32

5. Which fact regarding the periodic table is false?

    A. The elements are classified into groups by electronic structure

    B. The noble gases make up the most stable period

    C. Horizontal rows represent periods

    D. Elements in the same family have similar chemical properties

    E. Vertical rows represent groups

6. Both ^{65}Cu and ^{65}Zn have the same:

    A. mass number   **C.** number of ions

    B. number of neutrons   **D.** number of electrons   **E.** number of protons

7. Which group of elements contains only non-metals?

    A. Br, Ar, Na   **C.** Be, Ca, Sr

    B. S, As, Se   **D.** Fe, Ru, Mn   **E.** N, Se, Br

8. If one mole of an element weighs 12 grams, the mass number of this element has a value of approximately:

    A. 6.02×10^{23} **B.** 12   **C.** 1   **D.** $12 \times 6.02 \times 10^{23}$ **E.** 24

9. According to John Dalton, atoms of a given element:

    A. are divisible   **C.** are identical

    B. have the same shape   **D.** have different masses   **E.** none of the above

10. In which order must elements on the periodic table be listed for their properties to repeat at regular intervals?

    A. increasing atomic mass   **C.** increasing atomic number

    B. decreasing atomic mass   **D.** decreasing atomic number   **E.** decreasing density

11. Which of the following is a general characteristic of a metallic element?

A. high melting point   **C.** ductile

B. conductor of electricity   **D.** shiny luster   **E.** all of the above

12. Which of the following represents only alkali metals?

     I. Li   III. Rb

     II. Ba   IV. Ca

A. II and IV   **B.** I and III   **C.** III and IV   **D.** I and II   **E.** I and IV

13. Elements that appear in the same column of the periodic table and have similar chemical properties are called:

A. congeners   **C.** anomers

B. epimers   **D.** monomers   **E.** diastereomers

14. Which of the following is an alkali metal?

A. N   **B.** He   **C.** Cl   **D.** Ga   **E.** K

15. An element can be determined by:

A. A + C   **B.** C only   **C.** A only   **D.** Z only   **E.** none of the above

16. Based on the Law of Mass Conservation, Lavoisier hypothesized that:

A. an element is a combination of substances

B. carbon dioxide is a fundamental element

C. an element is made of a fundamental substance that cannot be broken down further

D. matter can lose mass as hot, dry, cold or moist qualities change

E. matter can gain mass as hot, dry, cold or moist qualities change

17. Elements constituting a period in the periodic table:

A. are always in the same group   **C.** are called isotopes

B. have consecutive atomic numbers   **D.** have similar chemical properties

                                      E. are called ions

18. Which of the following biochemical elements is classified as a trace element?

A. selenium   **B.** magnesium   **C.** potassium   **D.** chloride   **E.** oxygen

19. Which of the following is a general characteristic of a metallic element?

A. reacts with nonmetals   **C.** dull, brittle solid

B. reacts with metals   **D.** low melting point   **E.** none of the above

20. Which of the following gives the ground-state electron configuration of a Ca^{2+} atom?

A. [Ar] $4s^2\,3d^2$ **B.** [Ar] $4s^2\,4p^2$ **C.** [Ar]   **D.** [Ar] $4s^2$ **E.** [Ar] $4s^2\,4p^1$

Copyright © 2015 Sterling Test Prep

21. Which of the following represent(s) only halogens?

  I. O III. I

  II. He IV. Br

  A. II and III **B.** I and IV **C.** I and II **D.** III and IV **E.** I and IV

22. Sulfur has three common oxidation states: +2, 0, –2. Which oxidation state has the largest radius?

  A. anion **C.** neutral atom

  B. cation **D.** all are about the same size **E.** cation and anion

23. Which of the following substances does NOT contain cathode ray particles?

  A. H_2O **C.** hydrogen atoms

  B. Na **D.** all contain cathode ray particles **E.** He

24. Mg is an example of:

  A. a halogen **C.** an alkali metal

  B. a noble gas **D.** a transition metal **E.** an alkaline earth metal

25. Which has a mass of approximately 1 amu?

  A. 1 atom of ^{12}C **C.** 1 proton

  B. 1 mole of ^{12}C **D.** 1 electron **E.** 1 mole of electrons

26. Which of these does NOT describe a metal at room temperature?

  A. liquid **B.** shiny **C.** bendable **D.** solid **E.** gas

27. Which of the following sets of elements consists of members of the *same* group in the periodic table?

  A. ^{14}Si, ^{15}P and ^{16}S **C.** ^9F, ^{10}Ne and ^{11}Na

  B. ^{20}Ca, ^{26}Fe and ^{34}Se **D.** ^{31}Ga, ^{49}In and ^{81}Tl **E.** ^{11}Na, ^{20}Ca and ^{39}Y

28. Which of the following is a general characteristic of a nonmetal?

  A. low density **C.** brittle

  B. nonconductor of heat **D.** solid or gaseous state **E.** all of the above

29. Which of the following elements is a metal?

  A. Hydrogen **B.** Magnesium **C.** Silicon **D.** Chlorine **E.** Carbon

30. Which element has the greatest ionization energy?

  A. Al **B.** Cl **C.** P **D.** S **E.** Fr

31. What is the number of neutrons in an As isotope?

  A. 34 **B.** 36 **C.** 42 **D.** 40 **E.** 44

32. Which classification of the element is incorrect?

    A. Ce – Transition metal   **C.** Cd – Transition metal

    B. Ne – Noble gas   **D.** Li – Alkali metal   **E.** Br – Halogen

33. Which elements are most common in living organisms?

    A. O, Na, H, Ca   **C.** C, S, K, Mg

    B. H, O, N, C   **D.** Ca, O, S, Cl   **E.** H, Ca, Na, Cl

34. Which of the following is a general characteristic of a nonmetallic element?

    A. nonconductor of electricity   **C.** dull appearance

    B. low melting point   **D.** reactivity with metals   **E.** all of the above

35. Which of the following is the electron configuration of a boron atom?

    A. $1s^2\,2s^1\,2p^2$ **C.** $1s^2\,2s^2\,2p^2$

    B. $1s^2\,2p^3$ **D.** $1s^2\,2s^2\,2p^1$ **E.** $1s^2\,2s^1\,2p^1$

36. Which of the following elements is a nonmetal?

    A. Sodium   **B.** Chlorine   **C.** Aluminum   **D.** Magnesium   **E.** Palladium

37. Which element would most likely be a metal with a low melting point?

    A. K   **B.** B   **C.** N   **D.** C   **E.** Cl

38. Which of the following particles would NOT be deflected by charged plates?

    A. alpha particles   **C.** hydrogen atoms

    B. protons   **D.** cathode rays   **E.** All are deflected by charged plates

39. How many electrons can occupy the $n = 2$ shell?

    A. 2   **B.** 6   **C.** 8   **D.** 18   **E.** 32

40. How much does 4 moles of Na weigh?

    A. 5.75 grams   **B.** 23 amu   **C.** 23 grams   **D.** 92 grams   **E.** 92 amu

41. Which of these is NOT a metal?

    A. rubidium   **C.** aluminum

    B. sodium   **D.** vanadium   **E.** selenium

42. Which of the following elements occupies a position in Period 5 and Group IIIA on the periodic table?

    A. In   **B.** As   **C.** Tl   **D.** P   **E.** Sr

Copyright © 2015 Sterling Test Prep

43. Which of the following is a general characteristic of a nonmetallic element?

   A. reacts with nonmetals   **C.** pliable

   B. high melting point   **D.** shiny luster   **E.** none of the above

44. Which of the following elements is a metalloid?

   A. Copper   **B.** Iron   **C.** Silicon   **D.** Bromine   **E.** Palladium

45. Which characteristic is responsible for the changes seen in ionization energy when moving down a column?

   A. Increased shielding of electrons

   B. Increasing nuclear attraction for electrons and larger atomic or ionic radii

   C. Increasing nuclear attraction for electrons

   D. Decreasing nuclear attraction for electrons

   E. Increased shielding of electrons and larger atomic radii

46. How many electrons can occupy the $n = 4$ shell?

   A. 8   **B.** 10   **C.** 18   **D.** 32   **E.** 40

47. Which represents the charge on 1 mole of electrons?

   A. 6.02×10^{23} e   **C.** 6.02×10^{23} grams

   B. 6.02×10^{23} C   **D.** 1 C   **E.** 1 e

48. Which of these properties describes a metal?

   A. fragile   **C.** good conductor of heat

   B. transparent   **D.** brittle   **E.** poor conductor of electricity

49. The elements in groups IA, VIIA and VIIIA are called, respectively:

   A. alkaline earth metals, transition metals and halogens

   B. alkali metals, halogens and noble gases

   C. alkali metals, alkali earth metals and halogens

   D. alkaline earth metals, halogens and alkali metals

   E. halogens, alkali earth metals and noble gases

50. What does a positive charge on an atom signify?

   A. more protons than neutrons   **C.** more electrons than protons

   B. more neutrons than protons   **D.** more protons than electrons

                                      E. more electrons than neutrons

51. From the periodic table, which of the following elements is a semimetal?

   A. Ar   **B.** As   **C.** Al   **D.** Ac   **E.** Am

52. Which of the following is true of an element in an excited state?

    A. It has emitted a photon and its energy has decreased

    B. It has emitted a photon and its energy has increased

    C. It has absorbed a photon and its energy has decreased

    D. It has absorbed a neutron and its energy has increased

    E. It has absorbed a photon and its energy has increased

53. Which of the following represents a compound rather than an element?

    I. O_3 II. CCl_4 III. S_8 IV. H_2O

    A. I and III   **B.** II and IV   **C.** I and II   **D.** III and IV   **E.** I and IV

54. Which element has the lowest electronegativity?

    A. Mg   **B.** Al   **C.** Cl   **D.** Br   **E.** I

55. Uranium exists in nature in the form of several isotopes that have different:

    A. numbers of electrons   **C.** atomic numbers

    B. numbers of protons   **D.** charges   **E.** numbers of neutrons

56. How many electrons can occupy the $4s$ subshell?

    A. 1   **B.** 2   **C.** 6   **D.** 8   **E.** 10

57. The most common isotope of hydrogen contains how many neutrons?

    A. 0   **B.** 1   **C.** 2   **D.** 3   **E.** 4

58. Based on experimental evidence, Dalton postulated that:

    A. atoms of different elements have the same mass

    B. not all atoms of a given element are identical

    C. atoms can be created and destroyed in chemical reactions

    D. each element consists of indivisible minute particles called atoms

    E. none of the above

59. What is the maximum number of electrons in the n = 3 shell?

    A. 32   **B.** 18   **C.** 10   **D.** 8   **E.** 12

60. Referring to the periodic table, which of the following is a solid metal under normal conditions?

    A. Ce   **B.** Os   **C.** Ba   **D.** Cr   **E.** all of the above

61. What must be the same if two atoms represent the same element?

    A. number of neutrons   **C.** number of electron shells

    B. atomic mass   **D.** atomic number

                                     E. number of valence electrons

Copyright © 2015 Sterling Test Prep

62. Electron affinity is the:

   A. ability of an atom to attract electrons when it bonds with another atom

   B. energy needed to remove an electron from a neutral atom of the element in the gas phase

   C. energy liberated when an electron is added to a gaseous neutral atom converting it to an anion

   D. A and B

   E. A and C

63. How many electrons can occupy the $4d$ subshell?

   A. 2   **B.** 6   **C.** 8   **D.** 10   **E.** 12

64. Which statement is true regarding the relative abundances of the ^{6}lithium or ^{7}lithium isotopes?

   A. The relative proportions change as neutrons move between the nuclei

   B. The isotopes are in roughly equal proportions

   C. The relative ratio depends on the temperature of the element

   D. ^{6}Lithium is much more abundant

   E. ^{7}Lithium is much more abundant

65. An element in its ground state absorbs photons of 2.3 eV and 4.1 eV, but no intermediate energies. If the element in its ground state absorbs a photon of energy 4.1 eV, it emits a photon of 4.1 eV and a photon of:

   A. 1.8 eV   **B.** 6.4 eV   **C.** 2.3 eV   **D.** 4.1 eV   **E.** 0.8 eV

66. In the ground state of an atom:

   A. the energy of the electrons are at a maximum

   B. the protons and neutrons fill all the available energy states

   C. all the electrons are in their lowest energy levels

   D. the excited states are all filled

   E. none of the above

67. Each atom of an element has the same number of:

   A. electrons   **C.** photons

   B. neutrons   **D.** protons   **E.** protons and electrons

68. Refer to the periodic table and predict which of the following is a solid nonmetal under normal conditions.

   A. Cl   **B.** F   **C.** Se   **D.** As   **E.** Ar

69. Except for helium, the valence shell of electrons in the noble gases has which of the following electron configurations?

   A. ns^2np^2 **B.** ns^2np^4 **C.** ns^8np^2 **D.** ns^2np^8 **E.** ns^2np^6

70. Which particle is in the nucleus?

   A. protons and neutrons   **C.** protons only

   B. protons and electrons   **D.** neutrons only   **E.** neutrons and electrons

71. Which element would most likely be a metalloid?

   A. B   **B.** Mg   **C.** Cl   **D.** H   **E.** C

72. All isotopes of an element possess the same:

   A. number of electrons, atomic number and mass, but have nothing else in common

   B. atomic number and mass, but have nothing else in common

   C. chemical properties and mass, but have nothing else in common

   D. number of electrons, atomic number and chemical properties

   E. mass only

73. What is the maximum number of electrons that can occupy the 4f orbitals?

   A. 6   **B.** 8   **C.** 14   **D.** 16   **E.** 18

74. ^{63}Cu is 69% of the naturally occurring isotope of Cu. If only one other isotope is present for natural copper, what is it?

   A. ^{59}Cu **B.** ^{65}Cu **C.** ^{61}Cu **D.** ^{62}Cu **E.** ^{60}Cu

75. Which of the following statements best describes an element?

   A. has consistent physical properties

   B. consists of more than one type of atom

   C. consists of only one type of atom

   D. material that is pure

   E. material that has consistent chemical properties

76. Which of the following statements are true?

     I. The f subshell contains 7 orbitals

     II. The third energy shell (n=3) has no f orbitals

     III. There are ten d orbitals in the d subshell

     IV. The second energy shell contains only s and p orbitals

     V. A p subshell can accommodate a maximum of 6 electrons

   A. II and IV   **C.** I, II, IV and V

   B. I, II and IV   **D.** II, III and V   **E.** II, III and IV

77. What is the term for the value which indicates the number of protons for an atom of a given element?

   A. atomic mass   **C.** atomic notation

   B. mass number   **D.** atomic number   **E.** none of the above

Copyright © 2015 Sterling Test Prep

78. Almost all of the mass of an atom exists in its:

   A. electrons   **C.** outermost energy level

   B. nucleus   **D.** first energy level   **E.** valence electrons

79. Which phrase describes the alkali metals (group IA)?

   A. Form strong ionic bonds with nonmetals   **C.** More reactive than group IIA elements

   B. Little or no reaction with water   **D.** A and C

                                                    E. B and C

80. Which group contains only metalloids?

   A. Pm, Sm and Nd   **C.** Cu, Ag and Au

   B. Ar, Xe and Rn   **D.** Po, Cs and Ac   **E.** Si, B and Sb

81. What is the mass of one mole of glucose which has a molecular formula of $C_6H_{12}O_6$?

   A. 90 g   **B.** 132 g   **C.** 180 g   **D.** 194 g   **E.** 360 g

82. Metalloids are elements:

   A. larger than nonmetals

   B. found in asteroids

   C. smaller than metals

   D. that have some properties like metals and some like nonmetals

   E. that have properties different from either the metals or the nonmetals

83. Which of the following subshells is lowest in energy?

   A. *5f*   **B.** *4s*   **C.** *3d*   **D.** *6s*   **E.** *5p*

84. The nucleus of an atom is made up of:

   A. protons and neutrons   **C.** protons

   B. electrons and neutrons   **D.** electrons and protons   **E.** neutrons

85. What is the term for a broad uninterrupted band of radiant energy?

   A. ultraviolet spectrum   **C.** continuous spectrum

   B. visible spectrum   **D.** radiant energy spectrum

                                      E. none of the above

86. Compared to the electronegativity of iodine, the electronegativity of chlorine is:

   A. lower, because increased nuclear shielding results in a weaker pull on the valence electrons

   B. greater, because decreased nuclear shielding allows for a stronger pull on the valence electrons

   C. lower, because Cl has a smaller nuclear charge which results in a smaller force exerted on electrons

   D. greater, because Cl has a greater nuclear charge which results in a greater force exerted on electrons

   E. greater, because Cl has a greater electron charge which results in a greater force exerted on the nucleus

87. Cobalt is element 27. ^{60}Cobalt is used in the medical treatment of cancer. How many neutrons and protons are contained in the nucleus of the ^{60}Cobalt isotope?

A. 27 neutrons and 27 protons

B. 33 neutrons and 33 protons

C. 27 neutrons and 33 protons

D. 33 neutrons and 33 electrons

E. 33 neutrons and 27 protons

88. Metalloids:

    I. have some metallic and some nonmetallic properties

    II. may have low electrical conductivities

    III. contain elements in group IIIB

A. I and III only

B. II only

C. II and III only

D. I and II only

E. I, II and III

89. All atoms of a particular element have the same:

A. number of electrons, atomic number and chemical properties, but not necessarily the same mass

B. chemical properties and the same mass, but lack other similarities

C. number of electrons, the same atomic number, the same mass, but lack other similarities

D. mass and the same chemical properties, but lack other similarities

E. atomic number and the same mass, but lack other similarities

90. What is the atomic number and the mass number of ^{79}Br, respectively?

A. 35; 44   **B.** 44; 35   **C.** 35; 79   **D.** 79; 35   **E.** 35; 114

91. With what charge do alkaline earth metals form ions?

A. –2   **B.** –1   **C.** 0   **D.** +1   **E.** +2

92. Which of the following is a metalloid?

A. antimony (Sb)

B. uranium (U)

C. iodine (I)

D. zinc (Zn)

E. selenium (Se)

93. Which types of subshells are present for the n = 2 energy level?

A. *d*   **B.** *p*   **C.** *s*   **D.** both *d* and *p*   **E.** both *s* and *p*

94. What is the term for the shorthand description of the arrangement of electrons by sublevels according to increasing energy?

A. continuous spectrum

B. electron configuration

C. atomic notation

D. atomic number

E. none of the above

95. Adding one proton to the nucleus of an atom:

A. causes no change in the atomic number and decrease in the atomic mass

B. increases its atomic number by one unit, but does not change its atomic mass

C. increases its atomic mass by one unit, but does not change its atomic number

D. increases the atomic number and the mass number by one unit

E. causes no change in the atomic mass and decrease in the atomic number

Copyright © 2015 Sterling Test Prep

96. Which subshell has a principal quantum number of 4 and an angular momentum quantum number of 2?

    A. $4d$ **B.** $4f$ **C.** $2p$ **D.** $4s$ **E.** none of the above

97. Which characteristics describe the mass, charge and location of a proton, respectively?

    A. approximate mass 5×10^{-4} amu; charge +1; inside nucleus

    B. approximate mass 1 amu; charge 0; inside nucleus

    C. approximate mass 1 amu; charge +1; inside nucleus

    D. approximate mass 5×10^{-4} amu; charge −1; outside nucleus

    E. approximate mass 1 amu; charge +1; outside nucleus

98. Halogens form ions by:

    A. gaining two electrons   **C.** gaining one neutron

    B. gaining one electron   **D.** losing one electron   **E.** losing two electrons

99. Which statement about a neutron is FALSE?

    A. It is a nucleon

    B. It is often associated with protons

    C. It is more difficult to detect than a proton or an electron

    D. It is much more massive than an electron

    E. It has a charge equivalent but opposite to an electron

100. What is the maximum number of electrons that can be placed into the *f* subshell, *d* subshell and *p* subshell respectively?

    A. 14, 10 and 6   **C.** 10, 14 and 6

    B. 12, 10 and 6   **D.** 2, 12 and 20   **E.** 16, 8 and 2

101. How many neutrons are in a Beryllium atom with an atomic number of 4 and atomic mass of 9?

    A. 4   **B.** 5   **C.** 9   **D.** 16   **E.** 20

102. Classify the following three elements of S, As & Ga as a metal, metalloid, or nonmetal:

    A. As, metal; Ga, metalloid; S, nonmetal   **C.** Ga, metal; As, metalloid; S, nonmetal

    B. Ga, metal; S, metalloid; As, nonmetal   **D.** S, metal; Ga, metalloid; As, nonmetal

                                                              E. S, metal; As, metalloid; Ga, nonmetal

103. Which of the following elements has the greatest ionization energy?

    A. scandium   **C.** potassium

    B. titanium   **D.** calcium   **E.** magnesium

104. Isotopes have the:

   A. same number of neutrons but different number of electrons

   B. same number of protons but different number of neutrons

   C. same number of protons but different number of electrons

   D. none of the above

   E. all of the above

105. Which characteristic is NOT a property of the transition elements?

   A. Colored ions in solution   **C.** Multiple oxidation states

   B. Complex ions formed   **D.** Nonmetallic in character

                                  E. None of the above

106. Which species shown below has 24 neutrons?

   A. $^{52}_{24}Cr$ **B.** $^{55}_{25}Mn$ **C.** $^{24}_{12}Mg$ **D.** $^{51}_{23}V$ **E.** $^{45}_{21}Sc$

107. Hydrogen exists as three isotopes which have different numbers of:

   A. neutrons   **C.** protons

   B. electrons   **D.** charges   **E.** protons and neutrons

108. The elements of silver, iron, mercury and rhodium are:

   A. transduction metals   **C.** alkaline earth metals

   B. actinoids   **D.** transition metals   **E.** lanthinides

109. Which atom is largest?

   A. Li   **B.** Na   **C.** K   **D.** H   **E.** Rb

110. Rank the elements below in order of decreasing atomic radius:

   A. Al > P > Cl > Na > Mg   **C.** Mg > Na > P > Al > Cl

   B. Cl > Al > P > Na > Mg   **D.** Na > Mg > Al > P > Cl

                                     E. P > Al > Cl > Mg > Na

111. Which of the following produces the "atomic fingerprint" of an element?

   A. excited protons dropping to a lower energy level

   B. excited protons jumping to a higher energy level

   C. excited electrons dropping to a lower energy level

   D. excited electrons jumping to a higher energy level

   E. none of the above

112. How many neutrons are in a neutral atom of ^{40}Ar?

   A. 18   **B.** 22   **C.** 38   **D.** 40   **E.** 60

 Copyright © 2015 Sterling Test Prep

113. Paramagnetism, the ability to be pulled into a magnetic field, is demonstrated by:

- **A.** any substance containing unpaired electrons
- **B.** nonmetal elements that have unpaired *p* orbital electrons
- **C.** transition elements that have unpaired *d* orbital electrons
- **D.** nonmetal elements that have paired *p* orbital electrons
- **E.** any substance containing paired electrons

114. The number of neutrons in an atom is equal to:

- **A.** the mass number
- **B.** the atomic number
- **C.** mass number minus atomic number
- **D.** atomic number minus mass number
- **E.** mass number plus atomic number

115. The element calcium belongs to which family?

- **A.** representative elements
- **B.** transition metals
- **C.** alkali metals
- **D.** lanthanides
- **E.** alkaline earth metals

116. Which of the following elements would be shiny and flexible?

- **A.** bromine (Br)
- **B.** selenium (Se)
- **C.** helium (He)
- **D.** ruthenium (Ru)
- **E.** silicon (Si)

117. Which of the following subshell notations for electron occupancy is an impossibility?

- **A.** $4f^{11}$
- **B.** $2p^1$
- **C.** $5s^3$
- **D.** $4p^5$
- **E.** $3p^3$

118. How many protons, neutrons and electrons does a Platinum atom have respectively, if an uncharged atom of Platinum has an atomic number of 78 and an atomic mass of 195?

- **A.** 78, 156, 78
- **B.** 117, 273, 117
- **C.** 117, 78, 117
- **D.** 117, 78, 78
- **E.** 78, 117, 78

119. Lines were observed in the spectrum of uranium ore identical to those of helium in the spectrum of the Sun. Which of the following produced the lines in the helium spectrum?

- **A.** excited protons jumping to a higher energy level
- **B.** excited protons dropping to a lower energy level
- **C.** excited electrons jumping to a higher energy level
- **D.** excited electrons dropping to a lower energy level
- **E.** none of the above

120. Compared to the atomic radius of calcium, the atomic radius of gallium is:

- **A.** smaller because increased nuclear charge causes electrons to be held more tightly
- **B.** larger because its additional electrons increase the atomic volume
- **C.** smaller because gallium gives up more electrons, thereby decreasing its size
- **D.** larger because increased electron charge requires that the same force be distributed over a greater number of electrons
- **E.** larger because decreased electron charge requires that the same force be distributed over a smaller number of electrons

121. What is the average atomic mass of an element that contains three isotopes of 16.0 amu, 17.0 amu, and 18.0 amu with relative abundances of 20%, 50% and 30%, respectively?

    A. 16.9 amu   **C.** 17.2 amu

    B. 17.1 amu   **D.** 17.4 amu   **E.** 17.5 amu

122. Element X has an atomic number of 7 and an atomic mass of 13. Element X has:

    A. 6 neutrons   **C.** 6 protons

    B. 6 electrons   **D.** 13 electrons   **E.** 7 neutrons

123. The species shown below which has 24 electrons is:

    A. $^{52}_{24}Cr$ **B.** $^{55}_{25}Mn$ **C.** $^{24}_{12}Mg$ **D.** $^{45}_{21}Sc$ **E.** $^{51}_{23}V$

124. An atom that contains 47 protons, 47 electrons and 60 neutrons is an isotope of:

    A. Nd   **B.** Bh   **C.** Ag   **D.** Al   **E.** cannot be determined

125. Which element listed below has the greatest electronegativity?

    A. I   **B.** Fr   **C.** H   **D.** He   **E.** F

126. Oxygen atoms form H_2O molecules. Do oxygen atoms in O_2 and in H_2O have similar properties?

    A. No, compounds are uniquely different from the elements from which they are made

    B. No, but their similar properties are only a coincidence

    C. Yes, but it is only a coincidence that their properties are similar

    D. Yes, and this explains how fish are able to breathe water

    E. Yes, compounds and the elements they originate from are identical

127. In an atom with many electrons, which of the following orbitals would be highest in energy?

    A. $4f$ **B.** $4p$ **C.** $6d$ **D.** $7s$ **E.** $8g$

128. Which element has the electron configuration of $1s^2 2s^2 2p^6 3s^2$?

    A. Na   **B.** Si   **C.** Ca   **D.** Mg   **E.** none of the above

129. Which of the following represents a pair of isotopes?

    A. $^{32}_{16}S$, $^{32}_{16}S^{2-}$ **C.** $^{14}_{6}C$, $^{14}_{7}N$

    B. O_2, O_3 **D.** $^{1}_{1}H$, $^{2}_{1}H$ **E.** none of the above

130. The atomic mass of sodium is 22.989 g/mole. This atomic mass is not an integer because of:

    A. quanta   **C.** Pauli exclusion principle

    B. uneven density   **D.** spin direction   **E.** isotopes

Copyright © 2015 Sterling Test Prep

131. How many protons and neutrons are in ^{35}Cl, respectively?

   A. 35; 18   **B.** 18; 17   **C.** 17; 18   **D.** 17; 17   **E.** 35; 17

132. The element with the least electronegativity is:

   A. Cl   **B.** Fr   **C.** I   **D.** F   **E.** C

133. What is the name of the compound composed of CaCl$_2$?

   A. dichloromethane   **C.** carbon chloride

   B. dichlorocalcium   **D.** calcium chloride   **E.** dicalcium chloride

134. The electron configuration for manganese (Mn) is:

   A. $1s^2 2s^2 2p^6 3s^2 3p^6$ **C.** $1s^2 2s^2 2p^6 3s^2 3p^6 4s^2 3d^6$

   B. $1s^2 2s^2 2p^6 3s^2 3p^6 4s^2 3d^{10} 4p^1$ **D.** $1s^2 2s^2 2p^6 3s^2 3p^6 4s^2 3d^6$ **E.** $1s^2 2s^2 2p^6 3s^2 3p^6 4s^2 3d^5$

135. Different isotopes of one element have an equal number of [   ], but different number of [   ].

   A. protons, electrons   **C.** neutrons, protons

   B. neutrons, electrons   **D.** protons, neutrons   **E.** electrons, protons

136. Which of the following is the best description of the Bohr atom?

   A. sphere with a heavy, dense nucleus surrounded by electrons
   B. sphere with a heavy, dense nucleus encircled by electrons in orbits
   C. indivisible, indestructible particle
   D. homogeneous sphere of *plum pudding*
   E. sphere with a heavy, dense nucleus surrounded by electron orbitals

137. All of the following are examples of polar molecules, EXCEPT:

   A. H$_2$O   **B.** CCl$_4$ **C.** CH$_2$Cl$_2$ **D.** HF   **E.** CO

138. The element rhenium (Re) exists as two stable and eighteen unstable isotopes. ^{185}Rhenium has a nucleus that contains:

   A. 110 protons and 130 neutrons   **C.** 75 protons and 130 neutrons
   B. 130 protons and 75 neutrons   **D.** 75 protons and 75 neutrons
                                             E. 75 protons and 110 neutrons

139. Which is the principal quantum number?

   A. n **B.** m **C.** l **D.** s **E.** +1/2

140. The atomic mass of a naturally occurring isotope of iron is reported as 55.434 amu, which means that the average mass is:

   A. 55.434/1.0079 times greater than a ^{1}H atom
   B. 55.434/12.000 times greater than a ^{12}C atom
   C. 55.434 times greater than a ^{12}C atom
   D. 55.434 times greater than a ^{1}H atom
   E. 55.434/12.011 times greater than a ^{12}C atom

141. An atom containing 29 protons, 29 electrons and 34 neutrons has a mass number of:

    A. 5   **B.** 29   **C.** 34   **D.** 63   **E.** 92

142. Ionization energy is generally defined as the energy:

    A. released when an element forms ions within an aqueous solution
    B. necessary to add an electron to an element in its standard state
    C. necessary to remove an electron from an element in its liquid state
    D. absorbed when an element forms ions within an aqueous solution
    E. necessary to remove an electron from an element in its gaseous state

143. Which of the following physical properties is expected for krypton (Kr)?

    A. brittle   **C.** exists as a gas at room temperature
    B. shiny   **D.** hard   **E.** conducts electricity

144. Which element has the electron configuration $1s^2 2s^2 2p^6 3s^2 3p^6 4s^2 3d^{10} 4p^6 5s^2 4d^{10} 5p^2$?

    A. Sn   **B.** As   **C.** Pb   **D.** Sb   **E.** In

145. In its ground state, how many unpaired electrons does a sulfur atom have?

    A. 0   **B.** 1   **C.** 2   **D.** 3   **E.** 4

146. The masses on the periodic table are expressed in what units?

    A. picograms   **C.** Grams
    B. nanograms   **D.** micrograms   **E.** Amu's

147. The *l* quantum number refers to the electron's:

    A. angular momentum   **C.** spin
    B. magnetic orientation   **D.** energy level   **E.** shell size

148. Which characteristics describe the mass, charge and location of a neutron, respectively?

    A. approximate mass 1 amu; charge 0; inside nucleus
    B. approximate mass 5×10^{-4} amu; charge 0; inside nucleus
    C. approximate mass 1 amu; charge +1; inside nucleus
    D. approximate mass 1 amu; charge −1; inside nucleus
    E. approximate mass 5×10^{-4} amu; charge −1; outside nucleus

149. Which of the following elements most easily accepts an extra electron?

    A. He   **B.** Ca   **C.** Cl   **D.** Fr   **E.** Na

150. What happens to the properties of elements across any period of the periodic table?

    A. Elements tend to become more metallic because of the increase in atomic number
    B. The properties of the elements change gradually across any period of the periodic table
    C. The elements get much larger in size because of the addition of more protons and electrons
    D. All of the above are true
    E. None of the above is true

Copyright © 2015 Sterling Test Prep

151. How many electrons are there in the outermost shell and subshell, respectively, in an atom with the electron configuration $1s^2 2s^2 2p^6 3s^2 3p^6 4s^2 3d^{10} 4p^1$?

    A. 4, 1   **B.** 3, 1   **C.** 10, 1   **D.** 2, 2   **E.** 5,1

152. Compared to nonradioactive isotopes, radioactive isotopes:

    A. have a lower number of neutrons   **C.** have a higher number neutrons

    B. have a higher number of protons   **D.** are more stable

                                          E. are less stable

153. What terms refer to a vertical column in the periodic table of elements?

    A. period or series   **C.** group or series

    B. period or family   **D.** group or family   **E.** none of the above

154. A β– particle is emitted by a lithium nucleus. The resulting nucleus is an isotope of:

    A. helium   **B.** carbon   **C.** beryllium   **D.** boron   **E.** lithium

155. How many electrons are in the highest energy level of sulfur?

    A. 2   **B.** 4   **C.** 6   **D.** 8   **E.** 10

156. Which statement is NOT true?

    A. The first four subshells that correspond to l = 0, 1, 2, and 3 are s, p, d, and f

    B. In the shell n= 2, there is 1 subshell

    C. The values of l range from 0 to (n – 1)

    D. The numbers of possible values for l describes the number of subshells in a shell

    E. The spin can be either + ½ or – ½

157. Which of the following is NOT an alkali metal?

    A. Fr   **B.** Cs   **C.** Ca   **D.** Na   **E.** Rb

158. Which characteristics describe an electron?

    A. approximate mass 5×10^{-4} amu; charge –1; outside nucleus

    B. approximate mass 5×10^{-4} amu; charge 0; inside nucleus

    C. approximate mass 1 amu; charge –1; inside nucleus

    D. approximate mass 1 amu; charge +1; inside nucleus

    E. approximate mass 1 amu; charge 0; inside nucleus

159. In a bond between any two of the following atoms, the bonding electrons would be most strongly attracted to:

    A. I   **B.** He   **C.** Cs   **D.** Cl   **E.** Fr

160. Dmitri Mendeleev's chart of elements:

A. predicted the behavior of missing elements

B. developed the basis of our modern periodic table

C. predicted the existence of elements undiscovered at his time

D. placed elements with the same number of valence electrons in the same horizontal row

E. all of the above

161. Which element has the electron configuration $1s^2 2s^2 2p^6 3s^2 3p^6 4s^2 3d^{10} 4p^6 5s^2 4d^1$?

   A. Y   **B.** La   **C.** Si   **D.** Sc   **E.** Zr

162. Which of the following elements contains 6 valence electrons?

   A. S   **B.** Cl   **C.** Si   **D.** P   **E.** Ca^{2+}

163. The magnetic quantum number does NOT:

A. suggest that there are always odd numbers of orbitals in a subshell

B. have values which range from -1 to $+1$

C. give the particular orbitals in a subshell

D. describe the spin of the electron

E. none of the above

164. Write the formula for the ionic compound formed from magnesium and sulfur:

   A. Mg_2S **B.** Mg_3S_2 **C.** MgS_2 **D.** MgS_3 **E.** MgS

165. Which has the largest radius?

   A. Br^- **B.** K^+ **C.** Ar   **D.** Ca^{2+} **E.** Cl^-

166. Which pair of elements have the correct magnetic properties?

A. Ba: paramagnetic   Xe: diamagnetic

B. Ba: paramagnetic   Xe: paramagnetic

C. Ba: diamagnetic   Xe: diamagnetic

D. Ba: diamagnetic   Xe: paramagnetic

E. B and D are correct

167. What is the maximum number of electrons to fill the atom's second electron shell?

   A. 18   **B.** 2   **C.** 4   **D.** 12   **E.** 8

168. Approximately what percent of a ^{14}carbon sample remains after 2,000 years if ^{14}carbon has a half-life of 5,700 years?

   A. 20%   **B.** 32%   **C.** 52%   **D.** 78%   **E.** 92%

169. Which of the following is the sulfate ion?

   A. SO_4^{2-} **B.** S^{2-} **C.** CO_3^{2-} **D.** PO_4^{3-} **E.** S^-

Copyright © 2015 Sterling Test Prep

170. What is the value of quantum numbers *n* and *l* in the highest occupied orbital for the element carbon that has an atomic number of 6?

A. $n = 1, l = 1$ **C.** $n = 1, l = 2$

B. $n = 2, l = 1$ **D.** $n = 2, l = 2$ **E.** $n = 3, l = 3$

171. Which element is a halogen?

A. Os   **B.** I   **C.** O   **D.** Te   **E.** Se

172. Which of the following is the correct sequence of atomic radii from smallest to largest?

A. $Al \rightarrow S \rightarrow Al^{3+} \rightarrow S^{2-}$ **C.** $Al^{3+} \rightarrow Al \rightarrow S \rightarrow S^{2-}$

B. $S \rightarrow S^{2-} \rightarrow Al \rightarrow Al^{3+}$ **D.** $Al^{3+} \rightarrow S \rightarrow Al \rightarrow S^{2-}$ **E.** $Al^{3+} \rightarrow Al \rightarrow S^{2-} \rightarrow S$

173. Early investigators proposed that the ray of the cathode ray tube was due to the cathode because the ray:

A. could be diverted by a magnetic field

B. was not seen from the positively charged anode

C. was attracted to the electric plates that were positively charged

D. would change colors depending on the gas used within the tube

E. was observed in the presence or absence of a gas

174. Which of the following atoms is paramagnetic?

A. Zn   **B.** O   **C.** Ca   **D.** Ar   **E.** Xe

175. What is the electron configuration for an atom of silicon?

A. $1s^2 2s^2 2p^4$ **C.** $1s^2 2s^2 2p^6 3s^2 3p^2$

B. $1s^2 2s^2 2p^6 3s^2$ **D.** $1s^2 2s^2 2p^6 3s^2 3p^4$ **E.** $1s^2 2s^2 2p^6 3s^2 3p^4$

176. What is the atomic number of an element with an electron configuration: $1s^2$, $2s^2$, $2p^6$, $3s^2$, $3p^4$?

A. 16   **B.** 14   **C.** 12   **D.** 10   **E.** Not enough information is provided

177. The smallest amount of an element that retains that element's characteristics is the:

A. molecule   **B.** neutron   **C.** atom   **D.** electron   **E.** proton

178. The attraction of the nucleus on the outermost electron in an atom tends to:

A. decrease from right–left and bottom–top on the periodic table

B. decrease from left–right and bottom–top on the periodic table

C. decrease from left–right and top–bottom of the periodic table

D. increase from right–left and top–bottom on the periodic table

E. decrease from right–left and top–bottom on the periodic table

179. What is the number of unpaired electrons in the electron configuration of arsenic?

A. 1   **B.** 3   **C.** 2   **D.** 0   **E.** 4

180. How many electrons are in its outermost electron shell if Oxygen has an atomic number of 8?

   A. 2   **B.** 6   **C.** 8   **D.** 16   **E.** 18

181. What is the mass of the sample of oxygen that contains 2.5 moles of O_2 molecules?

   A. 25 g   **B.** 40 g   **C.** 60 g   **D.** 80 g   **E.** 88 g

182. Which of the following is the carbonate ion?

   A. CO_3^{2-} **B.** CO_3 **C.** CO_2 **D.** CO_3^- **E.** SO_4^{2-}

183. Which of the following is NOT implied by the spin quantum number?

   A. The two spinning electrons generate magnetic fields

   B. The values are $+\frac{1}{2}$ or $-\frac{1}{2}$ **D.** A and B

   C. Orbital electrons have opposite spins   **E.** None of the above

184. Which statement does NOT describe the noble gases?

   A. The heavier noble gases react with other elements

   B. They belong to group VIIIA (or 18)

   C. They contain at least one metalloid

   D. He, Ne, Ar, Kr, Xe, Rn and Uuo are part of the group

   E. They were once known as the inert gases

185. The greatest dipole moment within a bond is when:

   A. both bonding elements have low electronegativity

   B. one bonding element has high electronegativity and the other has low electronegativity

   C. both bonding elements have high electronegativity

   D. one bonding element has high electronegativity and the other has moderate electronegativity

   E. both bonding elements have moderate electronegativity

186. Which of the elements would be in the same group as the element whose electronic configuration is $1s^2 2s^2 2p^6 3s^2 3p^6 4s^1$?

   A. ^{18}Ar **B.** ^{34}Se **C.** ^{12}Mg **D.** ^{15}P **E.** ^3Li

187. Which of the following is the symbol for the chlorite ion?

   A. ClO_2^- **B.** ClO^- **C.** ClO_4^- **D.** ClO_3^- **E.** ClO_2

188. The f subshell contains:

   A. 3 orbitals   **B.** 5 orbitals   **C.** 7 orbitals   **D.** 9 orbitals   **E.** 12 orbitals

189. Another name for atomic mass unit (amu) is the:

   A. Dalton   **B.** Kekule   **C.** Kelvin   **D.** Avogadro   **E.** Mendeleev

190. How many quantum numbers are needed to describe a single electron in an atom?

   A. 1   **B.** 2   **C.** 3   **D.** 4   **E.** 5

Copyright © 2015 Sterling Test Prep

191. Early investigators proposed that the ray of the cathode tube was actually a negatively charged particle because the ray was:

A. not seen from the positively charged anode

B. diverted by a magnetic field

C. observed in the presence or absence of a gas

D. able to change colors depending on which gas was within the tube

E. attracted to positively charged electric plates

192. If an element has an electron configuration ending in $3p^4$, which statements about the element's electron configuration is NOT correct?

A. There are six electrons in the 3rd shell

B. Five different subshells contain electrons

C. There are eight electrons in the 2nd shell

D. The 3rd shell needs two more electrons to be completely filled

E. none of the above

193. An atom with an electrical charge is referred to as a(n):

A. compound   B. ion   C. radioactive   D. isotope   E. element

194. What is the total mass of the H atoms contained in 3 moles of glucose ($C_6H_{12}O_6$)?

A. 3g   B. 12g   C. 36g   D. 48g   E. 58g

195. Which orbital is NOT correctly matched with its shape?

A. *d*, spherically symmetrical   C. *s*, spherically symmetrical

B. *p*, dumbbell shaped   D. B and C   E. A and B

196. The transition metals occur in which period(s) on the periodic table?

A. 2   B. 3   C. 4   D. 1   E. all of the above

197. The shell level of an electron is defined by which quantum number?

A. electron spin quantum number

B. magnetic quantum number

C. azimuthal quantum number

D. principal quantum number

E. principal quantum number and electron spin quantum number

198. Which of the elements is an alkaline earth metal?

A. ^{54}Xe B. ^{34}Se C. ^{21}Sc D. ^{37}Rb E. ^{38}Sr

199. Which one of the following correctly represents the electron configuration of sulfur in an excited state?

A. $1s^2 2s^2 2p^6 3s^2 3p^4 4s^1$ C. $1s^2 2s^2 2p^6 3s^2 3p^5$

B. $1s^2 2s^2 2p^6 3s^2 3p^3 4s^1$ D. $1s^2 2s^2 2p^6 3s^2 3p^4$ E. $1s^2 2s^2 2p^6 3s^1 3p^5$

Copyright © 2015 Sterling Test Prep

200. Electrons fill up subshells in order of:

A. decreasing distance from the nucleus   **C.** increasing energy

B. increasing distance from the nucleus   **D.** B and C   **E.** A and C

201. Which set of quantum numbers describe the highest energy electron?

A. $n = 2$; $l = 1$; $m_l = 0$; $m_s = -1/2$ **C.** $n = 1$; $l = 1$; $m_l = 0$; $m_s = +1/2$

B. $n = 3$; $l = 2$; $m_l = 2$; $m_s = -1/2$ **D.** $n = 1$; $l = 0$; $m_l = 0$; $m_s = -1/2$

E. $n = 2$; $l = 2$; $m_l = 2$; $m_s = -1/2$

202. Which of the following elements is NOT correctly classified?

A. Mo – transition element   **C.** K – representative element

B. Sr – alkaline earth metal   **D.** Ar – noble gas   **E.** Po – halogen

203. Which of the following Dalton's proposals is still valid?

A. Compounds contain atoms in small whole number ratios

B. Atoms of different elements combine to form compounds   **D.** All of the above

C. An element is composed of tiny particles called atoms   **E.** None of the above

204. Why does a chlorine atom form anions more easily than cations?

A. Chlorine has a high electronegativity value

B. Chlorine has a large positive electron affinity

C. Chlorine donates one electron to complete its outer shell

D. Chlorine gains one electron to complete its outer shell

E. Chlorine has a low electronegativity value

205. Which principle or rule states that only two electrons can occupy an orbital?

A. Pauli exclusion principle   **C.** Heisenberg's uncertainty principle

B. Hund's rule   **D.** Newton's principle   **E.** None of the above

206. Arrange the following elements in order of increasing atomic radius: Sr, Rb, Sb, Te, In

A. Te < Sb < In < Sr < Rb   **C.** Te < Sb < In < Rb < Sr

B. In < Sb < Te < Sr < Rb   **D.** Rb < Sr < In < Sb < Te   **E.** In < Sb < Te < Rb < Sr

207. Which set of quantum numbers is possible?

A. $n = 1$; $l = 2$; $m_l = 3$; $m_s = -1/2$ **C.** $n = 2$; $l = 1$; $m_l = 2$; $m_s = -1/2$

B. $n = 4$; $l = 2$; $m_l = 2$; $m_s = -1/2$ **D.** $n = 3$; $l = 3$; $m_l = 2$; $m_s = -1/2$

E. $n = 2$; $l = 3$; $m_l = 2$; $m_s = -1/2$

208. Which of the following pairs has one metalloid element and one nonmetal element?

A. ^{82}Pb and ^{83}Bi   **C.** ^{51}Sb and ^{20}Ca

B. ^{19}K and ^{9}F   **D.** ^{3}As and ^{14}Si   **E.** ^{32}Ge and ^{9}F

209. Which of the following ions has the same charge as the hydroxide ion?

A. CO_3^{2-} **B.** PO_4^{3-} **C.** NO_3^{-} **D.** NH_4^{+} **E.** H_3O^{+}

Copyright © 2015 Sterling Test Prep

210. When an atom is most stable, how many electrons does it contain in its valence shell?

   A. 4   **B.** 6   **C.** 8   **D.** 10   **E.** 12

211. The shell with a principal quantum number 3 can accommodate how many electrons?

   A. 2   **B.** 3   **C.** 10   **D.** 18   **E.** 8

212. What is the number of known nonmetals relative to the number of metals?

   A. about two times greater   **C.** about four times less

   B. about fifty percent   **D.** very small in comparison

                                          E. about three times greater

213. Which type of subshell is filled by the distinguishing electron of an alkaline earth metal?

   A. *s*   **B.** *p*   **C.** *f*   **D.** *d*   **E.** both *s* and *p*

214. An excited state of an atom is represented by which electron configuration?

   A. $1s^2 2s^2 2p^6 3s^1$ **C.** $1s^2 2s^2 2p^1$

   B. $1s^2 2s^2 2p^6$ **D.** $1s^2 2s^2 3s^1$ **E.** $1s^2 2s^2 2p^6 3s^2$

215. An excited hydrogen atom emits a light spectrum of specific, characteristic wavelengths. The light spectrum is a result of:

   A. energy released as H atoms form H_2 molecules

   B. the light wavelengths which are not absorbed by valence electrons when white light passes through the sample

   C. particles being emitted as the hydrogen nuclei decay

   D. excited electrons being promoted to higher energy levels

   E. excited electrons dropping to lower energy levels

216. Which electron configuration depicts an excited state of potassium that has an atomic number of 19?

   A. $1s^2 2s^2 2p^6 3s^2 3p^6$ **C.** $1s^2 2s^2 2p^6 3s^2 3p^5 4s^2$

   B. $1s^2 2s^2 2p^6 3s^2 3p^6 4s^2$ **D.** $1s^2 2s^2 2p^6 3s^2 3p^7$

                                          E. $1s^2 2s^2 2p^6 3s^2 3p^2 4s^6$

217. Which statement is true of the energy levels for an electron in a hydrogen atom?

   A. The energy levels are identical to the levels in the He^+ ion

   B. The energy of each level can be computed from a known formula

   C. The distance between energy levels for $n = 1$ and $n = 2$ is the same as the distance between the $n = 3$ and $n = 4$ energy levels

   D. Since there is only one electron, the electron must be located in the lowest energy level

   E. The distance between the $n = 3$ and $n = 4$ energy levels in the same as the distance between the $n = 4$ and $n = 5$ energy levels

218. Ignoring hydrogen, which area of the periodic table contains both metals and nonmetals?

A. both *p* and *d* areas   **C.** *s* area only

B. *p* area only   **D.** both *s* and *p* areas   **E.** both *s* and *d* areas

219. When an excited electron returns to the ground state, it releases:

A. photons   **C.** beta particles

B. protons   **D.** alpha particles   **E.** gamma particles

220. Which of the following molecules does NOT exist?

A. OF_5 **B.** $NaLiCO_3$ **C.** ICl **D.** UF_6 **E.** all of the above exist

221. Which column of the periodic table has 3 nonmetal and 2 metalloid elements?

A. Group IIIA   **C.** Group VA

B. Group IVA   **D.** Group IA   **E.** Group VIA

222. Given that parent and daughter nuclei are isotopes of the same element, the ratio of α to β decay produced by the daughter must be:

A. 1 to 1   **B.** 1 to 2   **C.** 2 to 1   **D.** 2 to 3   **E.** 3 to 2

223. Which is the name of the elements that have properties between true metals and true nonmetals?

A. alkaline earth metals   **C.** nonmetals

B. metalloids   **D.** metals   **E.** halogens

224. An ion is represented by which of the following electron configurations?

  I. $1s^2 2s^2 2p^6$ II. $1s^2 2s^2 2p^6 3s^2$ III. $1s^2 2s^2 2p^3 3s^2 3p^6$

A. I only   **C.** II and III only

B. I and II only   **D.** I, II, and III   **E.** II only

225. Which subshell has the correct order of increasing energy?

A. 3*s*, 3*p*, 4*s*, 3*d*, 4*p*, 5*s*, 4*d*   **C.** 3*s*, 3*p*, 3*d*, 4*s*, 4*p*, 4*d*, 5*s*

B. 3*s*, 3*p*, 4*s*, 3*d*, 4*p*, 4*d*, 5*s*   **D.** 3*s*, 3*p*, 3*d*, 4*s*, 4*p*, 5*s*, 4*d*

E. 3*s*, 3*p*, 4*s*, 3*d*, 4*d*, 4*p*, 5*s*

226. Which of the following electron configurations represents an excited state of an atom?

A. $1s^2 2s^2 2p^6 3s^2 3d^3 3p^6 4s^2$ **C.** $1s^2 2s^2 2p^6 3s^2 3d^{10}$

B. $1s^2 2s^2 2p^6 3s^2 3p^6 3d^{10} 4s^2 4p^6$ **D.** $1s^2 2s^2 2p^6 3s^2 3p^6 4s^1$

E. $1s^2 2s^2 2p^6 3s^2 3p^6 3d^{10} 4s^2 4p^3$

Copyright © 2015 Sterling Test Prep

Chapter 2. BONDING

1. Based on the Lewis structure and formal charge considerations, how many resonance structures, if any, can be drawn for the PO_4^{3-} ion?

    A. original structure only   **B.** 2   **C.** 3   **D.** 4   **E.** 5

2. Which of the substances below would have the largest dipole?

    I. CO_2 II. SO_2^- III. H_2O

    A. I

    B. II

    C. III

    D. each dipole is the same magnitude

    E. none of the molecules has a dipole

3. What is the type of bond that forms between oppositely charged ions?

    A. dipole   **B.** covalent   **C.** London   **D.** induced dipole   **E.** ionic

4. How many valence electrons are in a chlorine atom and a chloride ion?

    A. 17 and 18, respectively

    B. 8 and 7, respectively

    C. 7 and 8, respectively

    D. 1 and 8, respectively

    E. none of the above

5. What is the formula of the sulfite ion?

    A. SO_3^{2-} **B.** S^{2-} **C.** HSO_4^{2-} **D.** SO_4^{2-} **E.** none of the above

6. Which of these molecules has trigonal pyramidal molecular geometry?

    A. SO_3 **B.** NF_3 **C.** AlF_3 **D.** BF_3 **E.** CH_4

7. The valence shell is the:

    A. innermost shell that is complete with electrons

    B. last partially filled orbital of an atom

    C. shell of electrons in an atom that is the least reactive

    D. outermost shell of electrons around an atom

    E. same as the orbital configuration

8. The metaphosphate ion, PO_3^-, is the structural analog of the NO_3^- ion with respect to the arrangement of the atoms. From the Lewis structure for the metaphosphate ion, what is the number of resonance structures for the Lewis structure?

    A. original structure only

    B. 2

    C. 3

    D. 4

    E. 5

Copyright © 2015 Sterling Test Prep

9. A characteristic of a cation is that:

    A. the number of neutrons is related to the number of electrons

    B. it has less electrons than protons

    C. it has equal numbers of electrons and protons

    D. it has less protons than electrons

    E. it has equal numbers of electrons and protons but a different number of neutrons

10. Which is sufficient for determining the molecular formula of a compound?

    A. The % by mass and the molecular weight of a compound

    B. The % by mass and the empirical formula of a compound

    C. The % by mass of a compound

    D. The molecular weight of a compound

    E. The empirical formula of a compound

11. An ionic bond forms between two atoms when:

    A. protons are transferred from the nucleus of the nonmetal to the nucleus of the metal

    B. each atom acquires a negative charge

    C. electron pairs are shared

    D. four electrons are shared

    E. electrons are transferred from metallic to nonmetallic atoms

12. Which of the following statements is true?

    A. $-\Delta H_f^o$ reactions produce stable compounds

    B. An endothermic reaction produces a fairly stable compound

    C. Bond formation tends to increase the potential energy of the atoms

    D. A compound with $-\Delta H_f^o$ is unstable as it decomposes into its elements

    E. A compound with $+\Delta H_f^o$ is stable as it decomposes into its elements

13. The property that describes the ease with which an atom loses an electron to form a positive ion is:

    A. electronegativity   **C.** hyperconjugation

    B. ionization energy   **D.** electron affinity   **E.** none of the above

14. For a representative element, valence electrons are electrons:

    A. located in the outermost orbital   **C.** that are located closest to the nucleus

    B. located in the outermost *d* subshell   **D.** occupying the outermost *s* and *p* orbitals

                                              E. located in the outermost *f* subshell

15. What bond forms between: $Na^+ + Cl^- \rightarrow NaCl$?

    A. dipole   **B.** van der Waals   **C.** covalent   **D.** hydrogen   **E.** ionic

16. What is the total number of valence electrons in a sulfite ion, SO_3^{2-}?

    A. 22   **B.** 24   **C.** 26   **D.** 34   **E.** none of the above

                                      Copyright © 2015 Sterling Test Prep

17. How many valence electrons are in the electron configuration of $1s^2 2s^2 2p^6 3s^2 3p^5$?

    A. 6   **B.** 7   **C.** 5   **D.** 2   **E.** 1

18. An ion is:

    A. a molecule such as galactose
    B. a substance formed by the combination of two elements
    C. an atom that has an electrical charge
    D. another term for an atom
    E. an element with differences in the number of neutrons

19. Covalent bonds:

        I. involve the sharing of electrons so each atom acquires a noble gas configuration
        II. can be either polar or nonpolar
        III. are stronger than ionic bonds

    A. I only   **B.** II only   **C.** III only   **D.** I and II only   **E.** I, II and III

20. How many valence electrons do the elements in groups IIA and VA of the periodic table have, respectively?

    A. 2 and 5   **B.** 3 and 4   **C.** 2 and 6   **D.** 2 and 2   **E.** 2 and 4

21. Based on the ΔH°_{f} data, which compound is the most stable?

    A. PH_3 (g), +5.3 kJ/mol   **C.** H_2S (g), –20.5 kJ/mol
    B. N_2H_4 (g), +94.4 kJ/mol   **D.** NH_3 (g), –46.5 kJ/mol
                            E. N_2O_4 (g), +9.8 kJ/mol

22. What is the valence shell electron configuration of the ion formed from a halogen?

    A. ns^2 **B.** $ns^2 np^2$ **C.** $ns^2 np^4$ **D.** $ns^2 np^8$ **E.** $ns^2 np^6$

23. A single covalent bond is formed by how many electrons?

    A. 1   **B.** 2   **C.** 3   **D.** 4   **E.** 0

24. What is the number of valence electrons in antimony (Sb)?

    A. 1   **B.** 2   **C.** 3   **D.** 4   **E.** 5

25. Which statement is true about the valence shell?

    A. The valence shell determines the electron dot structure
    B. The electron dot structure is made up of each of the valence shells
    C. The valence shell is the innermost shell
    D. The valence shell is usually the most unreactive shell
    E. None of the above

26. Which bond is formed by the equal sharing of electrons?

    A. covalent   **B.** dipole   **C.** ionic   **D.** London   **E.** van der Waals

Copyright © 2015 Sterling Test Prep

27. What is the number of valence electrons in tin (Sn)?

    A. 14   **B.** 8   **C.** 2   **D.** 4   **E.** 5

28. Which of the following occurs naturally as nonpolar diatomic molecules?

    A. sulfur   **C.** argon
    B. chlorine   **D.** all of the above   **E.** none of the above

29. Which of the following describes the orbital geometry of an sp^2 hybridized atom?

    A. bent   **C.** trigonal bipyramidal
    B. tetrahedral   **D.** linear   **E.** trigonal planar

30. An ion with an atomic number of 34 and 36 electrons has what charge?

    A. +2   **B.** ⁻36   **C.** +34   **D.** ⁻2   **E.** neutral

31. The "octet rule" relates to the number eight because:

    A. all orbitals can hold 8 electrons
    B. electron arrangements involving 8 valence electrons are extremely stable
    C. all atoms have 8 valence electrons
    D. only atoms with 8 valence electrons undergo a chemical reaction
    E. each element can accommodate a number of electrons as an integer of 8

32. Covalently bonded substances:

    A. have higher melting-point substances than ionic solids
    B. are found only in liquid and gas phases
    C. are good conductors of heat
    D. are good conductors of electricity
    E. are soft

33. Which of the following represents the breaking of a noncovalent interaction?

    A. Ionization of water   **C.** Hydrolysis of an ester
    B. Decomposition of hydrogen peroxide   **D.** Dissolving of salt crystals
    E. None of the above

34. Which of the following statements about noble gases is NOT correct?

    A. They have very stable electron arrangements
    B. They are the most reactive of all gases
    C. They exist in nature as individual atoms rather than molecular form
    D. They have 8 valence electrons
    E. They have a complete octet

35. Which species has eight valence electrons in the Lewis structure?

    A. S^{2-} **B.** Mg^+ **C.** F^+ **D.** Ar^+ **E.** Si

Copyright © 2015 Sterling Test Prep

36. The property that describes the energy released by a gas phase atom from adding an electron is:

   A. ionization energy   **C.** electron affinity

   B. electronegativity   **D.** hyperconjugation   **E.** none of the above

37. Which of the following must occur for an atom to obtain the noble gas configuration?

   A. lose, gain or share an electron   **C.** lose an electron

   B. lose or gain an electron   **D.** share an electron   **E.** share or gain an electron

38. As bond length between a pair of atoms increases, the bond strength:

   A. decreases and bond energy increases   **C.** remains constant as does bond energy

   B. increases and bond energy decreases   **D.** increases and bond energy increases

                                             E. decreases as does bond energy

39. Which of the following elements has two valence electrons?

   A. Ne   **B.** H   **C.** Mg   **D.** K   **E.** Li

40. In the process of forming sodium nitride (Na_3N), what happens to the electrons of each sodium atom and the electrons of each nitride atom, respectively?

   A. lose one; gain three   **C.** lose three; gain one

   B. lose three; gain three   **D.** lose one; gain two   **E.** lose two; gain three

41. How many covalent bonds can a neutral sulfur form?

   A. 0   **B.** 1   **C.** 2   **D.** 3   **E.** 4

42. Which of the following molecules can form hydrogen bonds?

   A. NH_3 **C.** HF

   B. H_2O **D.** all of the above   **E.** none of the above

43. Which of the following pairings of ions shown are NOT consistent with the formula?

   A. Co_2S_3 (Co^{3+} and S^{2-})   **C.** Na_3P (Na^+ and P^{3-})

   B. K_2O (K^+ and O^-)   **D.** BaF_2 (Ba^{2+} and F^-)   **E.** KCl (K^+ and Cl^-)

44. The term for a bond where the electrons are shared unequally:

   A. ionic   **C.** coordinate covalent

   B. nonpolar covalent   **D.** nonpolar ionic   **E.** polar covalent

45. Which atom would NOT be bound to hydrogen in a hydrogen bond?

   A. S   **B.** N   **C.** F   **D.** O   **E.** A and B

46. Which is the correct formula for the ionic compound formed between Ca and I?

   A. Ca_3I_2 **B.** Ca_2I_3 **C.** CaI_2 **D.** Ca_2I **E.** Ca_3I_5

47. Which species below has the least number of valence electrons in its Lewis symbol?

    A. S^{2-} **B.** Mg^{2+} **C.** Ar^+ **D.** Ga^+ **E.** F^-

48. Which has the greatest ionization energy?

    A. P   **B.** Al   **C.** Ne   **D.** Br   **E.** Ca

49. Which statement is NOT correct?

    A. Cations and anions combine in the simplest ratio which results in electrical neutrality

    B. The number of electrons lost by the cation equals the number of electrons gained by the anion in an ionic compound

    C. Ionic compounds may contain one metal and one nonmetal

    D. Ionic compounds may contain one metal and one halogen

    E. Formulas of ionic compounds are written with the anion first, followed by the cation

50. When a bond is broken, energy is:

    A. absorbed if the bond strength is negative   **C.** always released

    B. released if the bond strength is negative   **D.** always absorbed

    E. absorbed if the bond strength is positive

51. A dipole is a:

    A. nonpolar entity   **C.** separation of charges

    B. form of electronegativity   **D.** molecule with parallel bonds

    E. form of hyperconjugation

52. Which of the compounds is most likely ionic?

    A. N_2O_5 **C.** CBr_4

    B. $SrBr_2$ **D.** GaAs   **E.** CH_2Cl_2

53. What bond holds hydrogens and oxygen together within a water molecule?

    A. covalent   **C.** hydrophilic

    B. dipole   **D.** hydrogen   **E.** van der Waals

54. Which molecular geometry CANNOT result in a nonpolar structure?

    A. diatomic covalent   **C.** tetrahedral

    B. square planar   **D.** trigonal planar   **E.** bent

55. Which of the statements is an accurate description of the structure of the ionic compound NaCl?

    A. Alternating rows of Na^+ and Cl^- ions are present

    B. Each ion present is surrounded by six ions of opposite charge

    C. Alternating layers of Na and Cl atoms are present

    D. Alternating layers of Na^+ and Cl^- ions are present

    E. Repeating layers of Na^+ and Cl^- ions are present

Copyright © 2015 Sterling Test Prep

56. The term for a bond where the electrons are shared equally:

    A. coordinate covalent   **C.** polar covalent

    B. ionic   **D.** nonpolar covalent   **E.** nonpolar ionic

57. Given that the first ionization energy of cesium is +376 kJ/mol and the electron affinity of bromine is –325 kJ/mol, calculate ΔE for the reaction: $Cs\ (g) + Br\ (g) \rightarrow Cs^+\ (g) + Br^-\ (g)$

    A. +51 kJ/mol   **C.** +376 kJ/mol

    B. –701 kJ/mol   **D.** +701kJ/mol   **E.** –51 kJ/mol

58. What term best describes the smallest whole number repeating ratio of ions in an ionic compound?

    A. lattice   **C.** formula unit

    B. unit cell   **D.** covalent unit   **E.** ionic unit

59. The property defined as the energy required to remove one electron from an atom in the gaseous state is:

    A. electronegativity   **C.** electron affinity

    B. ionization energy   **D.** hyperconjugation   **E.** none of the above

60. What is the name for the force holding two atoms together in a chemical bond?

    A. gravitational force   **C.** weak hydrophobic force

    B. strong nuclear force   **D.** weak nuclear force   **E.** electrostatic force

61. A polyatomic ion:

    A. develops a charge as a result of the combination of two or more types of atoms

    B. does not bond further with other ions

    C. has a negative charge of less than –1

    D. contains both a metal and a nonmetal

    E. remains neutral from the combination of two or more types of atoms

62. Which of the following is the strongest form an interatomic attraction?

    A. dipole-induced dipole interaction   **C.** ion-dipole interaction

    B. dipole-dipole interaction   **D.** covalent bond

        E. induced dipole-induced dipole interaction

63. Water is a polar molecule because oxygen:

    A. is at one end with the hydrogens at the other end of the molecule

    B. has a partial negative charge while the hydrogens have a partial positive charge

    C. has a partial positive charge while the hydrogens have a partial negative charge

    D. is bonded between the two hydrogens

    E. attracts the hydrogen atoms

64. Which formula for an ionic compound is NOT correct as written?

    A. NH_4ClO_4 **B.** MgS **C.** KOH **D.** $Ca(SO_4)_2$ **E.** KF

65. Which of the following describes the attraction between two NH_3 molecules?

    A. nonpolar covalent bond   **C.** coordinate covalent bond

    B. polar covalent bond   **D.** hydrogen bond

                               E. none of the above

66. Which of the following molecules is polar?

    A. SO_3 **B.** SO_2 **C.** CO_2 **D.** CH_4 **E.** CCl_4

67. What is the chemical formula for a compound that contains K^+ and CO_3^{2-} ions?

    A. $K(CO_3)_3$ **B.** $K_3(CO_3)_2$ **C.** $K_3(CO_3)_3$ **D.** KCO_3 **E.** K_2CO_3

68. Why does H_2O have an unusually high boiling point compared to H_2S?

    A. Hydrogen bonding

    B. Van der Waals forces

    C. H_2O molecules pack more closely than H_2S

    D. Covalent bonds are stronger in H_2O

    E. This is a false statement, because H_2O has a similar boiling point to H_2S

69. Which is the most likely noncovalent interaction between an alcohol and a carboxylic acid?

    A. formation of an anhydride bond   **C.** dipole-charge interaction

    B. dipole-dipole interaction   **D.** charge-charge interaction

                               E. induced dipole-dipole interaction

70. Which formula for an ionic compound is NOT correct?

    A. $Al_2(CO_3)_3$ **B.** Li_2SO_4 **C.** Na_2S **D.** $MgHCO_3$ **E.** K_2O

71. Which of the following solids is likely to have the smallest exothermic lattice energy?

    A. Al_2O_3 **B.** $AlCl_3$ **C.** $NaCl$ **D.** LiF **E.** $CaCl_2$

72. A halogen is expected to have [   ] ionization energy and [   ] electron affinity?

    A. small; small   **C.** large; large

    B. small; large   **D.** large; small   **E.** none of the above

73. What is the total number of valence electrons in a molecule of SOF_2?

    A. 18   **B.** 20   **C.** 8   **D.** 19   **E.** 26

74. Two atoms are held together by a chemical bond because:

    A. atomic nuclei are attracted to the bonding electrons

    B. bonding electrons form an electrostatic cloud with the nuclei on the exterior

    C. atomic nuclei attract each other

    D. bonding electrons attract each other

    E. bonding electrons form an electrostatic cloud that contains both nuclei

                                               Copyright © 2015 Sterling Test Prep

75. Which of the following is the weakest form of interatomic attraction?

    A. dipole-dipole   **C.** covalent bond

    B. dipole-induced dipole   **D.** ion-dipole   **E.** induced dipole-induced dipole

76. What is the total number of electrons shown in a correctly written Lewis structure for OF_2?

    A. 20   **B.** 26   **C.** 32   **D.** 18   **E.** 22

77. What type of bond joins adjacent water molecules?

    A. dipole   **C.** hydrophilic

    B. hydrogen   **D.** covalent   **E.** induced dipole

78. From the electronegativities below, which covalent single bond is the most polar?

Element:	H	C	N	O
Electronegativity	2.1	2.5	3.0	3.5

    A. O–C   **B.** O–N   **C.** N–C   **D.** C–H   **E.** C–C

79. Which of the molecules contains a covalent triple bond?

    A. SO_2 **B.** HCN   **C.** Br_2 **D.** $C_2H_2Cl_2$ **E.** C_2H_6

80. Use the data to calculate the lattice energy of sodium chloride. $\Delta H_{lattice}$ equals:

$$Na\ (s) \rightarrow Na\ (g) \qquad\qquad \Delta H_1 = +108\ kJ$$

$$\tfrac{1}{2}Cl_2\ (g) \rightarrow Cl\ (g) \qquad\qquad \Delta H_2 = +120\ kJ$$

$$Na\ (g) \rightarrow Na\ (g) + e^- \qquad\qquad \Delta H_3 = +496\ kJ$$

$$Cl\ (g) + e^- \rightarrow Cl^-\ (g) \qquad\qquad \Delta H_4 = -349\ kJ$$

$$Na\ (s) + \tfrac{1}{2}Cl_2\ (g) \rightarrow NaCl\ (s) \qquad\qquad \Delta H^o_f = -411\ kJ$$

    A. –429 kJ/mol   **C.** –349 kJ/mol

    B. –760 kJ/mol   **D.** 786 kJ/mol   **E.** –786 kJ/mol

81. An alkaline earth element is expected to have [ ] ionization energy and [ ] electron affinity.

    A. small; large   **C.** large; small

    B. small; small   **D.** large; large   **E.** none of the above

82. Which of the statements concerning covalent double bonds is correct?

    A. They always involve the sharing of 2 electron pairs

    B. They occur only between atoms containing 4 valence electrons

    C. They are found only in molecules containing polyatomic ions

    D. They are found only in molecules containing carbon

    E. They occur only between atoms containing 8 valence electrons

83. The distance between two atomic nuclei in a chemical bond is determined by the:

    A. size of the valence electrons

    B. size of the nucleus

    C. size of the protons

    D. balance between the repulsion of the nuclei and the attraction of the nuclei for the bonding electrons

    E. size of the neutrons

84. Two molecules, X and Y, are not miscible and have very different physical properties. Molecule X boils at 70 °C and freezes at –20 °C, while molecule Y boils at 45 °C and freezes at –90 °C. Which molecule is likely to have the largest dipole?

    A. molecule X   **C.** not enough information to determine

    B. molecule Y   **D.** both have similar dipoles

    E. X and Y are the same molecule but with different physical properties

85. Which of the general statements regarding covalent bond characteristics is NOT correct?

    A. Triple bonds are less than three times stronger than single bonds

    B. Triple bonds are possible when 3 or more electrons are needed to complete an octet

    C. Triple bonds are stronger than double bonds

    D. Double bonds are stronger than single bonds

    E. Double bonding can occur with Group VIIA elements

86. Why are adjacent water molecules attracted to each other?

    A. Ionic bonding between the hydrogens of H_2O

    B. Covalent bonding between adjacent oxygen

    C. Electrostatic attraction between the H of one H_2O and the O of another

    D. Covalent bonding between the H of one H_2O and the O of another

    E. Electrostatic attraction between the O of one H_2O molecule and the O of another

87. Which of the following is true of a hydrogen bond?

    A. The bond length is longer than a covalent bond

    B. The bond energy is less than a covalent bond

    C. The bond is between H and O, N, or F

    D. The bond is between two polar molecules

    E. All of the above

88. When drawing a Lewis dot structure, pairs of electrons that are not between atoms but are used to fill the octet are:

    A. excess electrons   **C.** lone pairs

    B. filled shells   **D.** bonding pairs   **E.** not shown on the structure

89. Which atom is the most electronegative?

    A. P   **B.** Cl   **C.** Li   **D.** Cs   **E.** Fr

 Copyright © 2015 Sterling Test Prep

90. What is the name for the weak forces of attraction between nonpolar molecules due to temporary dipoles between adjacent nonpolar molecules?

A. van der Waals forces

B. hydrophobic forces

C. hydrogen bonding forces

D. nonpolar covalent forces

E. hydrophilic forces

91. Which bonding is NOT possible for a carbon atom that has four valence electrons?

A. 1 single bond and 1 triple bond

B. 1 double bond and 1 triple bond

C. 4 single bonds

D. 2 single bonds and 1 double bond

E. 2 double bonds

92. Which statement is true about the formation of the stable ionic compound NaCl?

A. NaCl is stable because it is formed from the combination of isolated gaseous ions

B. The first ionization energy of sodium contributes favorably to the overall formation of NaCl

C. The net absorption of 147 kJ/mol of energy for the formation of the Na^+ (*g*) and Cl^- (*g*) is responsible for the formation of the stable ionic compound

D. The release of energy as NaCl (*s*) forms leads to an overall increase in the potential energy

E. The lattice energy provides the necessary stabilization energy for the formation of NaCl

93. The statement that best describes the formation of an ionic compound is that electrons:

A. move freely among a network of nuclei in fixed positions

B. are shared between two atoms and discrete molecules are formed

C. are transferred from a non-metal to a metal, and the resulting charged particles form a crystalline network

D. are transferred from a metal to a non-metal, and the resulting charged particles form a crystalline network

E. do not permit each atom to achieve an octet

94. Nitrogen has five valence electrons; which of the following types of bonding is possible?

A. three single bonds

B. one triple bond

C. one single and one double bond

D. all of the above

E. none of the above

95. The bond energy of a carbon-carbon double bond, when compared to the energy of a carbon-carbon single bond, is:

A. twice the bond energy of a single bond

B. more than twice the bond energy of a single bond

C. less than the bond energy of a single bond

D. equal in bond energy to a single bond

E. greater in bond energy than a single bond, but less than twice the bond energy

96. Which of the following molecules would contain a dipole?

A. F–F   B. H–H   C. Cl–Cl   D. H–F   E. all of the above

97. Which of the following statements concerning coordinate covalent bonds is correct?

   A. Once formed, they are indistinguishable from any other covalent bond

   B. They are always single bonds

   C. One of the atoms involved must be a metal and the other a nonmetal

   D. Both atoms involved in the bond contribute an equal number of electrons to the bond

   E. The bond is formed between two Lewis bases

98. What is the name for the attraction between H_2O molecules?

   A. adhesion   **C.** cohesion

   B. polarity   **D.** van der Waals   **E.** hydrophilicity

99. The approximate bond angle between atoms of a trigonal planar molecule is:

   A. 109.5°   **B.** 180°   **C.** 120°   **D.** 90°   **E.** 104.5°

100. Which of the following is an example of a molecule that contains a coordinate covalent bond?

   A. CH_4 **B.** NH_3 **C.** NH_4^+ **D.** H_2O **E.** C_2H_4

101. For two ions with charges q_1 and q_2 that are separated by a distance r, the potential energy can be calculated from Coulomb's law: $E = q_1q_2/k \cdot r$

Calculate the energy released when 1 mole NaCl is formed, if the constant $k = 1.11 \times 10^{-10}$ C^2/J·m, the charge on $Na^+ = +e$, and the charge on $Cl^- = -e$ (Note: $e = 1.602 \times 10^{-19}$ C, and r = 282 pm)

   A. –494 kJ/mol   **C.** +91.2 kJ/mol

   B. –288 kJ/mol   **D.** +421 kJ/mol   **E.** –464 kJ/mol

102. Which one of the compounds below is most likely to be ionic?

   A. CBr_4 **B.** H_2O **C.** CH_2Cl_2 **D.** NO_2 **E.** $SrBr_2$

103. Calculate the total number of electrons in the Lewis structure for PH_4^+?

   A. 10   **B.** 12   **C.** 9   **D.** 8   **E.** 6

104. What is the mass percent of nitrogen in NO_2?

   A. 12.0%   **B.** 25.5%   **C.** 30.4%   **D.** 33.3%   **E.** 50.0%

105. How do the electron-dot structures of elements in the same group of the periodic table compare?

   A. The number of electrons in the electron-dot-structure equals the group number for each element of the group

   B. Elements of the same group have the same number of valence electrons

   C. The number of valence shell electrons increases by one for each element from the top to the bottom of the group

   D. The structures differ by exactly two electrons between vertically consecutive elements

   E. Elements of the same period have the same number of valence electrons

Copyright © 2015 Sterling Test Prep

106. What is the geometry of a molecule with three pairs of bonded electrons and one pair of nonbonded electrons around a central atom?

    A. octahedral   **C.** linear

    B. trigonal planar   **D.** tetrahedral   **E.** trigonal pyramidal

107. A water spider is able to walk on the surface of water because:

    A. hydrophilic bonds hold H_2O molecules together

    B. H_2O has strong covalent bonds within its molecules

    C. surface tension from the adhesive properties of H_2O

    D. surface tension from the cohesive properties of H_2O

    E. a water spider is less dense than H_2O

108. Under what conditions is graphite converted to diamond?

    A. low temperature, high pressure   **C.** high temperature, high pressure

    B. high temperature, low pressure   **D.** low temperature, low pressure

                                                      E. none of the above

109. What is the geometry of a molecule in which the central atom has 2 bonding electron pairs and 2 nonbonding electron pairs?

    A. trigonal planar   **C.** linear

    B. trigonal pyramidal   **D.** bent   **E.** tetrahedral

110. The approximate bond angle between atoms of a tetrahedral molecule is:

    A. 104.5°   **B.** 109.5°   **C.** 90°   **D.** 120°   **E.** 180°

111. Which force is intermolecular?

    A. polar covalent bond   **C.** dipole-dipole interactions

    B. ionic bond   **D.** nonpolar covalent bond   **E.** all are intermolecular forces

112. What is the molecular geometry of Cl_2CO?

    A. tetrahedral   **C.** trigonal pyramidal

    B. trigonal planar   **D.** linear   **E.** bent

113. Which statement is true?

    A. The buildup of electron density between two atoms repels each nucleus, making it less stable

    B. Bond energy is the minimum energy required to bring about the pairing of electrons in a covalent bond

    C. As the distance between the nuclei decreases for covalent bond formation, there is a corresponding decreased probability of finding both electrons near either nucleus

    D. One mole of hydrogen atoms is more stable than one mole of hydrogen molecules

    E. Two electrons in a single covalent bond are paired according to the Pauli exclusion principle

114. Which element forms an ion with the greatest positive charge?

    A. Mg   **B.** Sr   **C.** Al   **D.** Na   **E.** P

115. What is the molecular geometry of NCl_3?

   A. tetrahedral   **C.** trigonal bipyramidal

   B. trigonal planar   **D.** trigonal pyramidal   **E.** bent

116. Which statement describes the bond lengths in $CS_3{}^{2-}$?

   A. All bonds are of different lengths   **C.** Two are the same length while the other is longer

   B. All bonds are of the same length   **D.** Two are the same length while the other is shorter

                                               E. Depends if the molecule is a solid or liquid

117. How many more electrons can fit within the valence shell of a hydrogen atom?

   A. 1   **B.** 2   **C.** 7   **D.** 0   **E.** 3

118. Which one of the following molecules has a tetrahedral shape?

   A. NH_3 **B.** BF_3 **C.** C_2H_4 **D.** XeF_4 **E.** CF_4

119. Sweat cools the human body because water has:

   A. low specific heat   **C.** low density

   B. high specific heat   **D.** low molecular mass   **E.** high density

120. What is the molecular geometry of a $CH_3{}^+$ molecule?

   A. bent   **C.** tetrahedral

   B. trigonal pyramidal   **D.** linear   **E.** trigonal planar

121. What is the molecular geometry of PH_3?

   A. tetrahedral   **C.** trigonal pyramidal

   B. octahedral   **D.** linear   **E.** bent

122. Which bond is the most polar?

   A. H—O   **B.** H—P   **C.** H—N   **D.** H—C   **E.** H—Se

123. To form an octet, an atom of selenium must:

   A. gain 2 electrons   **C.** gain 6 electrons

   B. lose 2 electrons   **D.** lose 6 electrons   **E.** gain 4 electrons

124. What is the molecular geometry of CO_2?

   A. bent   **C.** trigonal planar

   B. tetrahedral   **D.** linear   **E.** trigonal bipyramidal

125. In which compound does carbon have the greatest percent by mass?

   A. $C_3H_7NH_2$ **B.** C_3H_7OH **C.** CH_3OH **D.** CCl_4 **E.** CH_2Cl_2

126. Which of the following has the greatest number of nonbonding pairs of electrons?

   A. He   **B.** F   **C.** C   **D.** H   **E.** S

Copyright © 2015 Sterling Test Prep

127. Which of the following pairs is NOT correctly matched?

Formula   Molecular Geometry

   A. CH_4 tetrahedral
   B. OF_2 bent
   C. PCl_3 trigonal planar
   D. Cl_2CO trigonal planar
   E. $^{\oplus}CH_3$ trigonal planar

128. What is the molecular geometry of a NH_3 molecule?

   A. bent   **C.** tetrahedral
   B. trigonal pyramidal   **D.** trigonal planar   **E.** linear

129. The strongest intermolecular force between molecules is:

   A. hydrophobic   **C.** polar
   B. Van der Waals   **D.** coordinate covalent bonds   **E.** hydrogen bonding

130. Electronegativity is a concept that is used to:

   A. formulate a statement of the octet rule
   B. evaluate if double bonds are present in a molecule
   C. evaluate how many electrons are involved in bonding
   D. evaluate the polarity of a bond
   E. evaluate if electrons are moved further from the nucleus of the electronegative element

131. Arrange the following in increasing electronegativity:

   A. Ga < Ge < As < P   **C.** P < Ga < Ge < As
   B. Ge < P < As < Ga   **D.** As < P < Ga < Ge   **E.** Ga < P < As < Ge

132. The charge on a sulfide ion is:

   A. +3   **B.** +2   **C.** 0   **D.** –2   **E.** –3

133. The ability of an atom in a molecule to attract electrons to itself is:

   A. ionization energy   **C.** electronegativity
   B. paramagnetism   **D.** electron affinity   **E.** hyperconjugation

134. Which of the following compounds is NOT possible?

   A. H^- because H forms only positive ions
   B. PbO_2 because the charge on a Pb ion is only +2
   C. SF_6 because F does not have an empty *d* orbital to form an expanded octet
   D. CH_3^- because C lacks *d* orbitals
   E. OCl_6 because O does not have *d* orbitals to form an expanded octet

135. What is the molecular geometry of CH_2O?

   A. trigonal planar   **C.** linear
   B. tetrahedral   **D.** trigonal pyramidal   **E.** bent

136. In the nitrogen monoxide molecule, the dipole moment is 0.16 D and the bond length is 115 pm. What is the sign and magnitude of the charge on the oxygen atom:

   A. −0.098 *e*   **B.** −0.71 *e*   **C.** −1.3 *e*   **D.** −0.029 *e*   **E.** +1.3 *e*

137. Which of the following series of elements are arranged in the order of increasing electronegativity?

   A. Fr, Mg, Si, O   **C.** F, B, O, Li
   B. Br, Cl, S, P   **D.** Cl, S, Se, Te   **E.** Br, Mg, Si, N

138. The name of Cl^- is:

   A. chlorite ion   **C.** chloride ion
   B. chlorate ion   **D.** chlorine ion   **E.** diatomic chlorine

139. What is the percent by mass of Cl in carbon tetrachloride?

   A. 25%   **B.** 66%   **C.** 78%   **D.** 92%   **E.** 33%

140. Which statement about electronegativity is NOT correct?

   A. Electronegativity increases from left to right within a row
   B. Electronegativity increases from bottom to top within a group
   C. Fluorine is the most electronegative atom of all the elements
   D. Francium is the least electronegative element
   E. Metals generally have higher electronegativity values than nonmetals

141. The number of unpaired valence electrons for an atom is related to the number of bonds that the atom can form because the number of unpaired valence electrons:

   A. is twice the number of bonds that the atom can form
   B. is the same as the number of bonds that the atom can form
   C. has no defined relationship between the number of unpaired valence electrons and number of bonds that the atom can form
   D. is one-half the number of bonds that the atom can form
   E. is not related to the number of bonds that the atom can form

142. Which compound is covalent?

   A. RbF   **B.** HF   **C.** NaF   **D.** LiF   **E.** NaCl

143. Which atom is the least electronegative?

   A. Rb   **B.** Ca   **C.** F   **D.** Si   **E.** I

144. The stability of an ionic crystal is greatest by forming:

   A. strong intramolecular forces
   B. the most symmetrical arrangement of anions and cations
   C. the closest arrangement of anions and cations while minimizing repulsion forces
   D. the most complex arrangement of anions and cations while minimizing repulsion forces
   E. greater distant between anions and cations

          Copyright © 2015 Sterling Test Prep

145. Based on the Lewis structure, how many non-bonding electrons are on N in the nitrate ion?

    A. 0   **B.** 2   **C.** 4   **D.** 6   **E.** 8

146. A bond where the electrons are shared unequally is:

    A. ionic   **C.** coordinate covalent

    B. nonpolar covalent   **D.** van der Waals force   **E.** polar covalent

147. The name of S^{2-} is:

    A. sulfite ion   **C.** sulfur

    B. sulfide ion   **D.** sulfate ion   **E.** sulfurous ion

148. What is the empirical formula of the compound with the mass percent of 71.65% Cl, 24.27% C and 4.07% H?

    A. ClC_2H_5 **B.** Cl_2CH_2 **C.** $ClCH_3$ **D.** $ClCH_2$ **E.** CCl_4

149. In chlorine monoxide, chlorine has a charge of $+0.167\ e^-$. If the bond length is 154.6 pm, the dipole moment of the molecule is?

    A. 2.30 D   **B.** 0.167 D   **C.** 1.24 D   **D.** 3.11 D   **E.** 1.65 D

150. Which of the following diatomic molecules contains the bond of greatest polarity?

    A. CH_4 **B.** BrI   **C.** Cl–F   **D.** P_4 **E.** Te–F

151. Which element likely forms a cation with a +2 charge?

    A. K   **B.** S   **C.** Si   **D.** Mg   **E.** Br

152. The mass percent of a compound is as follows: 71.65% Cl, 24.27% C and 4.07% H. If the molecular weight of the compound is 98.96, what is the molecular formula of the compound?

    A. $Cl_2C_3H_6$ **B.** $Cl_2C_2H_4$ **C.** $ClCH_3$ **D.** ClC_2H_2 **E.** CCl_4

153. Based on the Lewis structure for hydrogen peroxide, H_2O_2, how many polar bonds and nonpolar bonds are present?

    A. 3 polar bonds and zero nonpolar bonds   **C.** 1 polar bond and 2 nonpolar bonds

    B. 2 polar bonds and 2 nonpolar bonds   **D.** zero polar bonds and 3 nonpolar bonds

                                          E. 2 polar bonds and 1 nonpolar bond

154. Which compound contains only covalent bonds?

    A. $HC_2H_3O_2$ **B.** NaCl   **C.** NH_4OH **D.** $Ca_3(PO_4)_2$ **E.** LiF

155. Which pair of elements is most likely to form an ionic compound when reacted together?

    A. C and F   **C.** Al and Si

    B. K and Br   **D.** Fe and Ca   **E.** H and N

156. What is the empirical formula of acetic acid, CH_3COOH?

    A. CH_3COOH **B.** $C_2H_4O_2$ **C.** CH_2O **D.** COH_2 **E.** CHO

157. Which of the following pairs is NOT correctly matched?

    <u>Formula</u>   <u>Molecular polarity</u>

  A. SiF_4 nonpolar

  B. H_2O polar

  C. HCN   nonpolar

  D. H_2CO polar

  E. CH_2Cl_2 polar

158. Based on the Lewis structure for H_3C-NH_2, the formal charge on N is:

  A. –2   **B.** 0   **C.** –3   **D.** +2   **E.** +3

159. What is the formula of the ammonium ion?

  A. NH_4^- **B.** N_4H^+ **C.** Am^+ **D.** Am^- **E.** NH_4^+

160. What is the empirical formula of the compound that has a mass percent of 6% H and 94% O?

  A. HO   **B.** H_2O **C.** H_2O_2 **D.** H_3O_3 **E.** H_4O_2

161. Which of the following pairs is NOT correctly matched?

    <u>Formula</u>   <u>Molecular polarity</u>

  A. PF_3 polar

  B. BCl_3 nonpolar

  C. H_2O polar

  D. BeF_2 polar

  E. CCl_4 nonpolar

162. Based on the Lewis structure, how many polar and nonpolar bonds are present for H_2CO?

  A. 1 polar bond and 2 nonpolar bonds   **C.** 3 polar bonds and zero nonpolar bonds

  B. 2 polar bonds and 1 nonpolar bond   **D.** zero polar bonds and 3 nonpolar bonds

                                           E. 2 polar bonds and 2 nonpolar bonds

163. What is the formula of the carbonate ion?

  A. $C_2O_4^{2-}$ **B.** $C_2O_4^{1-}$ **C.** CO_2^{3-} **D.** CO_3^{2-} **E.** $C_2H_3O_2^{1-}$

164. Which of the following molecules is polar?

  A. CBr_4 **B.** C_2Cl_6 **C.** $BeCl_2$ **D.** CCl_4 **E.** NCl_3

165. How many total resonance structures, if any, can be drawn for a nitrite ion?

  A. original structure only   **B.** 2   **C.** 3   **D.** 4   **E.** 5

166. What is the formula of a nitrate ion?

  A. NO_3^{1-} **B.** NO_2^{1-} **C.** NO^{3-} **D.** NO_3^{2-} **E.** none of the above

                                               Copyright © 2015 Sterling Test Prep

Chapter 3. STATES OF MATTER: GASES, LIQUIDS, SOLIDS

1. Which of the following laws states that for a gas at constant temperature, the pressure and volume are inversely proportional?

A. Dalton's law
C. Boyle's law
B. Gay-Lussac's law
D. Charles's law
E. none of the above

2. What is standard atmospheric pressure of 760 mmHg in inches Hg? (2.54 cm = 1 in)

A. 73.00 in Hg
C. 1,930 in Hg
B. 29.92 in Hg
D. 103 in Hg
E. 109 in Hg

3. 212 °F is equivalent to how many degrees Kelvin?
A. 473   B. 278   C. 273   D. 493   E. 373

4. Which characteristics best describe a gas?

A. volume and shape of container; no intermolecular attractions
B. definite volume; shape of container; weak intermolecular attractions
C. definite shape and volume; strong intermolecular attractions
D. definite volume; shape of container; moderate intermolecular attractions
E. volume and shape of container; strong intermolecular attractions

5. Which of the following statements about gases is correct?

A. formation of homogeneous mixtures regardless of the natures of non-reacting gas components
B. relatively long distances between molecules
C. high compressibility
D. no attractive forces between gas molecules
E. all of the above

6. Which of the following laws states that the volume and Kelvin temperature are directly proportional for a gas at constant pressure?

A. Gay-Lussac's law
C. Charles's law
B. Dalton's law
D. Boyle's law
E. none of the above

7. What is the name given to the transition from the gas phase directly to the solid phase?

A. deposition
C. freezing
B. sublimation
D. condensation
E. evaporation

8. At what temperature is degrees Celsius equivalent to degrees Fahrenheit?
A. 0   B. –20   C. –40   D. –60   E. 10

9. A closed-end manometer was constructed from a U-shaped glass tube. It was loaded with mercury so that the closed side was filled to the top, which was 800 mm above the neck while the open end was 180 mm above the neck. The manometer was taken into a chamber used for training astronauts. What is the highest pressure that can be read with assurance on this manometer?

    A. 62.0 torr   **B.** 620 torr   **C.** 98.0 torr   **D.** 98.5 torr   **E.** 6.20 torr

10. Which characteristics best describe a liquid?

    A. volume and shape of container; no intermolecular attractions
    B. definite volume; shape of container; no intermolecular attractions
    C. definite shape and volume; strong intermolecular attractions
    D. definite volume; shape of container; moderate intermolecular attractions
    E. volume and shape of container; strong intermolecular attractions

11. Gas laws describe in mathematical terms the relationships between pressure and which two variables for a fixed quantity of gas.

    A. chemical identity; cost   **C.** temperature; volume
    B. volume; chemical identity   **D.** temperature; size   **E.** volume; size

12. Which of the following laws states that the pressure exerted by a gas is inversely proportional to its volume and directly proportional to its Kelvin temperature?

    A. combined gas law   **C.** Boyle's law
    B. Gay-Lussac's law   **D.** Charles's law   **E.** none of the above

13. Which of these alkanes has the lowest boiling point?

    A. C_8H_{18} **B.** C_6H_{14} **C.** C_4H_{10} **D.** C_2H_6 **E.** $C_{10}H_{22}$

14. What are the units of the gas constant R?

    A. atm·K/l·mol   **C.** mol·l/atm·K
    B. atm·K/mol·l   **D.** mol·K/l·atm   **E.** l·atm/mol·K

15. Which is NOT a correct statement of Boyle's Law?

    A. $P_1V_1 = P_2V_2$
    B. pressure × volume = a constant
    C. pressure α 1/volume
    D. a gas sample at constant temperature, pressure and volume are inversely proportional
    E. none of the above

16. Which of the following is NOT a unit used in measuring pressure?

    A. kilometer Hg   **C.** atmosphere
    B. millimeters Hg   **D.** inches Hg   **E.** torr

Copyright © 2015 Sterling Test Prep

17. Which of the following laws states that the pressure exerted by a mixture of gases is equal to the sum of the individual gas pressures?

  A. Gay-Lussac's law   **C.** Charles's law

  B. Dalton's law   **D.** Boyle's law   **E.** none of the above

18. Which of the following compounds will have the lowest boiling point?

  A. CH_4 **B.** $CHCl_3$ **C.** CH_3CH_2OH **D.** NH_3 **E.** CH_2Cl_2

19. What is the volume percent of Ar in a 6.50 L flask that contains 0.200 mole of Ne, 0.300 mole He and 0.600 mole of Ar at STP?

  A. 44.3%   **B.** 21.9%   **C.** 36.4%   **D.** 54.5%   **E.** 16.8 %

20. A U.S. Weather Bureau forecast cited the atmospheric pressure at sea level as 768.2 mm Hg. What is this value expressed in kilopascals (kPa)?

(1 atm = 101325 Pa = 760 torr = 760 mm Hg = 1.01325 bar = 1013.25mb)

  A. 778.4 kPa   **C.** 100.3 kPa

  B. 1024 kPa   **D.** 91.62 kPa   **E.** 102.4 kPa

21. What is the proportionality of volume and pressure for a gas?

  A. directly   **C.** raised to the 2nd power

  B. inversely   **D.** none of the above   **E.** all of the above

22. When volatile solvents X and Y are mixed in equal proportions, heat is released to the surroundings. If pure X has a higher boiling point than pure Y, which is NOT true?

  A. The vapor pressure of the mixture is less than pure Y

  B. The vapor pressure of the mixture is less than pure X

  C. The boiling point of the mixture is less than pure X

  D. The boiling point of the mixture is less than pure Y

  E. Not enough information is provided

23. How, in terms of size, is mmHg of pressure units related to the torr units?

  A. 1000 times larger than   **C.** equal to

  B. 760 times larger than   **D.** 100 times larger than   **E.** 760 times smaller than

24. What is the term that refers to the frequency and energy of gas molecules colliding with the walls of the container?

  A. partial pressure   **C.** atmospheric pressure

  B. vapor pressure   **D.** gas pressure   **E.** none of the above

25. Which of the following compounds has the highest boiling point?

  A. $CH_3CH_2CH_3$ **C.** CH_3CH_3

  B. $CH_3CH_2CH_2CH_3$ **D.** CH_4 **E.** CCl_4

26. 190 torr is equivalent to how many atmospheres?

    A. 0.19   **B.** 0.25   **C.** 0.30   **D.** 1.9   **E.** 3.8

27. Which characteristics best describe a solid?

    A. definite volume; shape of container; no intermolecular attractions
    B. volume and shape of container; no intermolecular attractions
    C. definite shape and volume; strong intermolecular attractions
    D. definite volume; shape of container; moderate intermolecular attractions
    E. volume and shape of container; strong intermolecular attractions

28. What is standard atmospheric pressure in inches mercury (in Hg) where 2.54 cm = 1 in?

    A. 1930 in Hg   **C.** 76.0 in Hg
    B. 101 in Hg   **D.** 88 in Hg   **E.** 29.9 in Hg

29. Which of the following laws states that the pressure and Kelvin temperature are directly proportional for a gas at constant volume?

    A. Gay-Lussac's law   **C.** Charles's law
    B. Dalton's law   **D.** Boyle's law   **E.** none of the above

30. Which of the following alkanes has the highest boiling point?

    A. $CH_3C(CH_3)_2CH_2CH_3$ **C.** $CH_3CH_2CH_2CH_2CH_2CH_2CH_3$
    B. $CH_3CH(CH_3)CH_2CH_2CH_3$ **D.** $CH_3CH(CH_3)CH(CH_3)CH_3$
                                     E. all have equivalent boiling points

31. Under which conditions does a real gas behave most nearly like an ideal gas?

    A. high temperature and high pressure   **C.** low temperature and low pressure
    B. high temperature and low pressure   **D.** low temperature and high pressure
    E. if it remains in the gaseous state regardless of temperature or pressure

32. If an ionic bond is stronger than a dipole-dipole interaction, why does water dissolve an ionic compound?

    A. Ions do not overcome their interatomic attraction and therefore are not soluble
    B. Ion-dipole interaction causes the ions to heat up and vibrate free of the crystal
    C. Ionic bond is weakened by the ion-dipole interactions and ionic repulsion ejects the ions
        from the crystal
    D. Ion-dipole interactions of several of water molecules aggregate with the ionic bond and
        dissociate it into the solution
    E. none of the above

33. The barometric pressure in the eye of a hurricane dips as low as 27.2 inches Hg. How many millimeters of Hg is this where 2.54 cm = 1 in?

    A. 107 mmHg   **C.** 6.91 mmHg
    B. 691 mmHg   **D.** 1.07 mmHg   **E.** 0.691 mmHg

Copyright © 2015 Sterling Test Prep

34. Which of the following increases the pressure of a gas?

    A. decreasing the volume

    B. increasing the number of molecules

    C. increasing temperature

    D. none of the above

    E. all of the above

35. Which of the following is true when comparing two compounds of similar molar mass in which compound A is comprised of nonpolar molecules while compound B is composed of polar molecules?

    A. B boils at a higher temperature than A

    B. B boils at a lower temperature than A

    C. A does not boil

    D. B does not boil

    E. Both compounds have the same boiling point

36. At the molecular level, ideal gases:

    A. do not occupy space

    B. occupy space

    C. do not exhibit intramolecular forces

    D. exhibit intramolecular forces

    E. A and C

37. A sample of a gas in a cylindrical chamber with a movable piston occupied a volume of 4.626 liters when the pressure was 0.983 atm and the temperature was 27.2 °C. By moving the piston, the pressure was adjusted to 1.388 atm. What is the volume occupied under the new conditions if the temperature remains constant?

    A. 0.303 L   **B.** 3.28 L   **C.** 4.68 L   **D.** 6.35 L   **E.** 6.49 L

38. In comparing gases to liquids, gases have:

    A. smaller compressibility and greater density

    B. smaller compressibility and smaller density

    C. greater compressibility and greater density

    D. greater compressibility and smaller density

    E. none of the above

39. 8 liters of O_2 gas is under a pressure of 280 torr. If the volume of this gas is increased to 14 liters at constant temperature, what is the new pressure?

    A. 390 torr   **B.** 160 torr   **C.** 66 torr   **D.** 490 torr   **E.** 6 torr

40. A sample of krypton gas at 75.0 psi and 100 °C expands from 0.100 L to 0.450 L. What is the final pressure in psi when the temperature remains constant?

    A. 16.7 psi   **B.** 75.0 psi   **C.** 1.67 psi   **D.** 3.38 psi   **E.** 33.8 psi

41. Which compound would have the highest boiling point?

    A. CH_3OH

    B. $CH_3CH_2CH_2CH_2CH_2-OH$

    C. $CH_3-O-CH_2\ CH_2CH_2CH_3$

    D. $CH_3CH_2-O-CH_2\ CH_2CH_3$

    E. $CH_3CH_2CH_2C(OH)H-OH$

42. Which transformation describes evaporation?

A. solid → liquid   **C.** liquid → solid

B. solid → gas   **D.** liquid → gas   **E.** gas → solid

43. The kinetic-molecular theory of ideal gases assumes which of the following?

        I. Gas molecules move at the same speed
        II. Gas molecules have negligible volume
        III. Gas molecules exert no attractive forces on each other

A. I only   **C.** II and III only

B. I and III only   **D.** I, II and III   **E.** III only

44. What is the relationship between the pressure and volume of a fixed amount of gas at constant temperature?

A. directly proportional   **C.** inversely proportional

B. equal   **D.** decreased by a factor of 2   **E.** none of the above

45. Assuming constant pressure, if a volume of nitrogen gas at 420 K decreases from 100 mL to 50 mL, what is the final temperature in Kelvin?

A. 630 K   **B.** 420 K   **C.** 910 K   **D.** 150 K   **E.** 210 K

46. The major intermolecular force in $(CH_3)_2NH$ is:

A. hydrogen bonding   **C.** London-dispersion forces

B. dipole-dipole attractions   **D.** ion-dipole attractions   **E.** van der Waals forces

47. Which is an assumption of the kinetic molecular theory of gases?

A. Nonelastic collisions   **C.** Nonrandom collisions

B. Constant interaction of molecules   **D.** Gas particles take up space

                                            E. Elastic collisions

48. A cylinder fitted with a movable piston and filled with a gas has a volume of 16.44 liters at 22.4 °C at an applied pressure of 772.2 torr. The temperature of the surrounding oil bath was increased to 184.4 °C while the load on the piston was changed. The volume was 16.60 liters. What is the final pressure in the system?

A. 496 torr   **B.** 504.2 torr   **C.** 1,184 torr   **D.** 1,209 torr   **E.** 6,288 torr

49. Which transformation describes condensation?

A. solid → gas   **C.** liquid → gas

B. solid → liquid   **D.** gas → liquid   **E.** liquid → solid

50. What property primarily determines the effect of temperature on the solubility of gas molecules?

A. ionic strength of the gas   **C.** polarity of the gas

B. molecular weight of the gas   **D.** kinetic energy of the gas

                                          E. dipole strength of the solvent

 Copyright © 2015 Sterling Test Prep

51. A 2.50 L sample of He gas has a pressure of 0.925 atm. What is the pressure of the gas if the volume is reduced to 0.350 L?

    A. 6.61 atm   **C.** 0.130 atm

    B. 0.661 atm   **D.** 0.946 atm   **E.** 13.0 atm

52. A sample of air at 7.50 atm is heated from 220 K to 440 K when the volume remains constant, what is the final pressure?

    A. 30.0 atm   **C.** 4.57 atm

    B. 6.15 atm   **D.** 3.75 atm   **E.** 15.0 atm

53. The attraction due to London dispersion forces between molecules depends on what two factors?

    A. Volatility and shape   **C.** Vapor pressure and size

    B. Molar mass and volatility   **D.** Molar mass and shape

                                   E. Molar mass and vapor pressure

54. What is the new internal pressure of a given mass of nitrogen gas in a 400-ml vessel at 22 °C and 1.2 atm is heated to 60 °C and compressed to 300 ml?

    A. (1.2)(0.300)(333) / (1)(0.400)(295)

    B. (1)(0.400)(333) / (1.2)(0.300)(295)

    C. (1.2)(0.400)(295) / (1)(0.300)(333)

    D. (1.2)(0.400)(0.300) / (1)(333)(278)

    E. (1.2)(0.400)(333) / (1)(0.300)(295)

55. The standard reference conditions for gases are:

    A. temperature: 0.00 K; pressure: 1.000 standard atmosphere

    B. temperature: 0.00 °C; pressure: 1.000 standard atmosphere

    C. temperature: 273.15 K; pressure: 1.000 Pascal

    D. temperature: 298.15 K; pressure: 1.000 standard atmosphere

    E. temperature: 298.15 K; pressure: 1.000 Pascal

56. Which transformation describes freezing?

    A. solid → liquid   **C.** liquid → solid

    B. solid → gas   **D.** liquid → gas   **E.** gas → liquid

57. In the ideal gas law, what does *V* represent?

    A. Volume of a gas molecule

    B. Volume of the container which encloses the gas

    C. Average speed of a gas molecule

    D. Average velocity of a gas molecule

    E. Average kinetic energy of a gas molecule

Copyright © 2015 Sterling Test Prep

58. What happens if the pressure of a gas above a liquid increases such as by pressing a piston above a liquid?

    A. Pressure goes down and the gas moves out of the solvent

    B. Pressure goes down and the gas goes into the solvent

    C. The gas is forced into solution and the solubility increases

    D. The solution is compressed and the gas is forced out of the solvent

    E. The amount of gas in the solution remains constant

59. A mathematical statement of Charles' law is:

    A. $V_1 T_1 = V_2 T_2$ **C.** $V_1 + T_2 = V_2 + T_1$

    B. $V_1 + T_1 = V_2 + T_2$ **D.** $V_1 / T_2 = T_1 / V_2$ **E.** $V_1 / T_1 = V_2 / T_2$

60. If the temperature of a liquid increases, what happens to its vapor pressure?

    A. unpredictable   **C.** decreases

    B. remains constant   **D.** increases   **E.** none of the above

61. Which of the following statements about intermolecular forces is true?

    A. Hydrogen bonding occurs between any two molecules that contain hydrogen atoms

    B. Dipole-dipole interactions occurs between two polar molecules

    C. London dispersions forces are the strongest of the three types

    D. Intermolecular forces occur within molecules rather than between the molecules

    E. Intermolecular forces are of about the same magnitude

62. How many moles of carbon atoms are contained in a 22.4 liter sample of a gas at STP that contains, by volume, 20% C_2H_6, 50% CH_4 and 30% N_2?

    A. 0.40   **B.** 0.50   **C.** 0.90   **D.** 1.20   **E.** 0.75

63. Which transformation describes melting?

    A. solid → liquid   **C.** liquid → solid

    B. solid → gas   **D.** liquid → gas   **E.** gas → liquid

64. Which of the following statements best gives the kinetic molecular theory explanation of the Boyle's law observation of increased pressure due to decreased volume?

    A. Particles strike container walls with more force   **C.** Particles have more kinetic energy

    B. Particles strike container walls more frequently   **D.** Particles increase in size

    E. Particles have less kinetic energy

65. If the temperature of a liquid decreases, what happens to vapor pressure?

    A. remains constant   **C.** increases

    B. unpredictable   **D.** decreases   **E.** none of the above

66. When NaCl dissolves in water, the force of attraction that exists between Na^+ and H_2O is:

    A. van der Waals   **C.** ion-ion

    B. hydrogen bonding   **D.** dipole-dipole   **E.** ion-dipole

Copyright © 2015 Sterling Test Prep

67. A sample of an unknown gas was isolated in a gas containment bulb on a manifold. The volume of the bulb was 1.425 liters, the temperature was 25.40 °C, and the manifold pressure was 583.0 torr. What is the volume of this gas at STP?

    A. 1.000 L   **B.** 1.149 L   **C.** 1.670 L   **D.** 2.026 L   **E.** 11.38 L

68. Charles' law involves which of the following?

    A. indirect proportion   **C.** constant volume

    B. varying mass of gas   **D.** varying temperature   **E.** constant temperature

69. What is the vapor pressure of water at 100 °C?

    A. 760 mm Hg   **C.** 76 mm Hg

    B. 100 mm Hg   **D.** 1 mm Hg   **E.** none of the above

70. Which type of attractive forces occur in all molecules regardless of the atoms they possess?

    A. Dipole–ion interactions   **C.** Dipole–dipole attractions

    B. Ion–ion interactions   **D.** Hydrogen bonding

    E. London dispersion forces

71. One liter of an ideal gas is placed in a piston at 27 °C. If the pressure is constant and the temperature is changed to 50 K, the final volume is:

    A. 108 ml   **C.** 136 ml

    B. 131 ml   **D.** 167 ml   **E.** 184 ml

72. A sample of a gas occupying a volume of 122.4 ml at STP was placed in a different vessel with a volume of 164.2 ml, in which the pressure was measured as 0.9915 atm. What was its temperature?

    A. 90.2 °C   **C.** 124.1 °C

    B. 93.1 °C   **D.** 203.6 °C   **E.** 208.3 °C

73. Which transformation describes sublimation?

    A. solid → liquid   **C.** liquid → solid

    B. solid → gas   **D.** liquid → gas   **E.** gas → liquid

74. What is the relationship between temperature and volume of a fixed amount of gas at constant pressure?

    A. equal   **C.** directly proportional

    B. indirectly proportional   **D.** decreased by a factor of 2

    E. none of the above

75. A beaker of water at 22 °C is placed in a closed container and a vacuum pump is used to evacuate the air in the container. Why does the water begin to boil?

    A. The vapor pressure decreases   **C.** The atmospheric pressure is reduced

    B. Air is released from the water   **D.** The vapor pressure increases

    E. none of the above

76. Which quantity contains the greatest number of moles?

A. 10g SiO_2 C. 10g CBr_4

B. 10g SO_2 D. 10g CO_2 E. 10g CH_4

77. How does a real gas deviate from an ideal gas?

I. Molecules occupy a significant amount of space

II. Intermolecular forces may exist

III. Pressure is created from molecular collisions with the walls of the container

A. I only   C. I and II only

B. II only   D. II and III only   E. I and III only

78. What would be the new pressure if a 400 mL gas sample at 360 mm Hg is expanded to 800 mL with no change in temperature?

A. 180 mm Hg   C. 720 mm Hg

B. 360 mm Hg   D. 760 mm Hg   E. 800 mm Hg

79. 10.0 liters of O_2 gas are at a temperature of 23 °C. If the temperature of the gas is raised to 40 °C at constant pressure the new volume is:

A. 10.6 liters   C. 4.20 liters

B. 14.1 liters   D. 1.82 liters   E. 6.80 liters

80. If hydrogen gas is collected over water at 20 °C and 763 mm Hg, what is the partial pressure of the H_2? The vapor pressure of water at 20 °C is 17.5 mm Hg.

A. 763 mm Hg   C. 743 mm Hg

B. 745.5 mm Hg   D. 17.5 mm Hg   E. 780.5 mm Hg

81. What is the percent by mass of salt in a mixture that contains 150 g of salt, 1.2 kg of flour and 650 g of sugar?

A. 0.085%   C. 9.1%

B. 7.5%   D. 15%   E. 18%

82. A balloon originally had a volume of 4.39 L at 44 °C and a pressure of 729 torr. At constant pressure, to what temperature (in °C) must the balloon be cooled to reduce its volume to 3.78 L?

A. 73 °C   B. 273 °C   C. 38 °C   D. 0 °C   E. 88 °C

83. If oxygen gas is collected over water at 25 °C and 775 torr, what is the partial pressure of the O_2? The vapor pressure of water at 25 °C is 23.8 torr.

A. 751 torr   B. 725 torr   C. 23.8 torr   D. 750 torr   E. 799 torr

84. Which of the following statements bests describes a liquid?

A. Definite shape but indefinite volume   C. Indefinite shape and volume

B. Indefinite shape but definite volume   D. Definite shape and volume

E. Definite shape but indefinite mass

          Copyright © 2015 Sterling Test Prep

85. What is the ratio of the diffusion rate of O_2 molecules to the diffusion rate of H_2 molecules when five moles of O_2 gas and five moles of H_2 gas are placed in a large vessel, such that the gases and vessel are at the same temperature?

    A. 4:1   **B.** 1:4   **C.** 12:1   **D.** 1:1   **E.** 2:1

86. Measured at STP, how many liters of pure oxygen gas are required for the complete combustion of 11.2 L of methane gas?

    A. 12.3 L   **B.** 13.2 L   **C.** 22.4 L   **D.** 31.6 L   **E.** 35.8 L

87. Which statement of Gay–Lussac's Law describes the behavior of a fixed amount of gas?

    A. As temperature increases, pressure decreases at constant volume

    B. As temperature increases, pressure increases at constant volume

    C. As pressure increases, volume increases at constant temperature

    D. As pressure increases, volume decreases at constant temperature

    E. As temperature increases, volume increases at constant pressure

88. Which affects the average force / unit area exerted by a gas on the wall of its container?

    I. Average speed of a gas molecule

    II. Frequency of collision between gas molecules and the wall of the container

    III. Volume of a gas molecule

    A. I only   **B.** I and III only   **C.** II and III only   **D.** I, II, and III   **E.** II only

89. What is the difference between a dipole-dipole and an ion–dipole interaction?

    A. one involves dipole attraction between neutral molecules while the other involves dipole interactions with ions

    B. one involves ionic molecules interacting with other ionic molecules while the other deals with polar molecules

    C. one involves salts and water while the other doesn't involve water

    D. one involves hydrogen bonding while the other does not

    E. none of the above

90. The Gay–Lussac's law of increased pressure due to increased temperature is explained using kinetic molecular theory in the following way: The pressure must increase because the:

    A. molecules increase in size   **C.** molecules decrease in size

    B. molecules move slower   **D.** molecules strike the container walls less often

                                        E. molecules strike the container walls more often

91. Which of the following is true of an ideal gas according to the kinetic theory?

    A. All molecular collisions have the same energy

    B. All molecules have the same kinetic energy   **D.** None of the above

    C. All molecules have the same velocity   **E.** All of the above

92. Which of the following statements best describes a solid?

    A. Indefinite shape but definite volume   **C.** Definite shape and volume

    B. Definite shape but indefinite volume   **D.** Indefinite shape and volume

                                                  E. Definite shape but indefinite mass

93. A vessel contains 32 g of CH_4 gas and 8.5 g of NH_3 gas at a combined pressure of 1.8 atm. What is the partial pressure of the NH_3 gas?

    A. 0.30 atm   **B.** 0.22 atm   **C.** 0.44 atm   **D.** 0.38 atm   **E.** 0.36 atm

94. A mathematical statement of Gay-Lussac's law is:

    A. $P_1 + T_2 = P_2 + T_1$ **C.** $P_1 / T_1 = P_2 / T_2$

    B. $P_1 T_1 = P_2 T_2$ **D.** $P_1 / T_1 = P_2 / T_1$ **E.** $P_1 + T_1 = P_2 + T_2$

95. What are the conditions for a real gas to behave most like an ideal gas?

    A. low temperature, high pressure   **C.** high temperature, high pressure

    B. low temperature, low pressure   **D.** high temperature, low pressure

                                                  E. none of the above

96. Which of the following statements best describes a gas?

    A. Definite shape but indefinite volume   **C.** Indefinite shape and volume

    B. Indefinite shape but definite volume   **D.** Definite shape and volume

                                                  E. Definite shape but indefinite mass

97. A sample of gas has a volume of 130 mL at 0.900 atm. What would be the volume if the pressure is decreased to 0.300 atm while temperature is held constant?

    A. 65 mL   **B.** 190 mL   **C.** 130 mL   **D.** 180 mL   **E.** 390 mL

98. If the pressure of a gas sample is doubled, according to Gay–Lussac's law, the gas sample:

    A. Kelvin is doubled   **C.** volume is doubled

    B. Kelvin decreases by a factor of 2   **D.** volume decreases by a factor of 2

                                                  E. volume decreases by a factor of 4

99. According to Boyle's law, what happens to a gas as the volume increases?

    A. The temperature increases   **C.** The pressure increases

    B. The temperature decreases   **D.** The pressure decreases   **E.** none of the above

100. Matter is nearly incompressible in which of these states?

    A. solid   **B.** liquid   **C.** gas   **D.** liquid and gas   **E.** solid and liquid

101. How many molecules of neon gas are present in 5 liters at 10°C and 300 mm Hg? R=0.0821

    A. $(300/760)(0.821)(5) / (6 \times 10^{23})(283)$ **C.** $(300)(5)(283)(6 \times 10^{23})$

    B. $(300/760)(5)(6 \times 10^{23}) / (0.0821)(283)$ **D.** $(300/760)(5)(283)(6 \times 10^{23}) / (0.821)$

                                                  E. $(300/283)(5)(6 \times 10^{23}) / (0.0821)(760)$

Copyright © 2015 Sterling Test Prep

102. A chemical reaction, A (*s*) → B (*s*) + C (*g*) occurs when substance *A* is vigorously heated. The molecular mass of the gaseous product was determined from the following experimental data:

Mass of *A* before reaction: 4.962 g

Mass of *A* after reaction: 0 g

Mass of residue *B* after cooling and weighing when no more gas was evolved: 3.684 g

When all of the gas *C* evolved, it was collected and stored in a 658.5 mL glass vessel at 30.4 °C and the gas exerted a pressure of 748.5 torr.

From this data, determine the apparent molecular mass of *C*, assuming it behaves as an ideal gas:

A. 6.46 g mol^{1-} **B.** 46.3 g mol^{1-} **C.** 49.1 g mol^{1-} **D.** 72.2 g mol^{1-} **E.** 142 g mol^{1-}

103. What is the new pressure if a sealed container with gas at 3.00 atm is heated from 30 K to 60 K?

A. 0.500 atm   **C.** 4.50 atm
B. 1.50 atm   **D.** 4.00 atm   **E.** 6.00 atm

104. Which of the following is a true statement when vessel I contains a gas at 200 °C while vessel II has the same gas at 100 °C.?

A. Each gas molecules in vessel I has more mass than each gas molecules in vessel II
B. All of the gas molecules in vessel I move slower than all of the gas molecules in vessel II
C. All of the gas molecules in vessel I move faster than all of the gas molecules in vessel II
D. None of the above statements are true
E. Both A and C are correct

105. The combined gas law can NOT be written as:

A. $V_2 = V_1 \times P_1 / P_2 \times T_2 / T_1$ **C.** $T_2 = T_1 \times P_1 / P_2 \times V_2 / V_1$
B. $P_1 = P_2 \times V_2 / V_1 \times T_1 / T_2$ **D.** $V1 = V_2 \times P_2 / P_1 \times T_1 / T_2$
                              E. none of the above

106. According to Charles's law, what happens to a gas as temperature increases?

A. The volume decreases   **C.** The pressure decreases
B. The volume increases   **D.** The pressure increases   **E.** none of the above

107. Which of the following compounds does NOT exhibit hydrogen bonding?

A. CH_4 **B.** HF   **C.** NH_3 **D.** H_2O **E.** CH_3OH

108. The Van der Waals equation is used to describe nonideal gases. The terms n^2a / v^2 and nb stand for, respectively: Van der Waals equation: $(P + n^2a / v^2)(V - nb) = nRT$

A. volume of gas molecules and intermolecular forces
B. nonrandom movement and intermolecular forces between gas molecules
C. nonelastic collisions and volume of gas molecules
D. intermolecular forces and volume of gas molecules
E. nonrandom movement and volume of gas molecules

Copyright © 2015 Sterling Test Prep

109. How does the volume of a fixed sample of gas change if both the pressure and Kelvin are doubled?

    A. decreases by a factor of 2   **C.** doubles

    B. increases by a factor of 4   **D.** does not change

    E. change cannot be determined without more information

110. A hydrogen bond is a special type of:

    A. dipole–dipole attraction involving hydrogen bonded to another hydrogen atom

    B. attraction involving any molecules that contain hydrogens

    C. dipole-dipole attraction involving hydrogen bonded to a highly electronegative atom

    D. dipole–dipole attraction involving hydrogen bonded to any other atom

    E. none of the above

111. If both the pressure and the temperature of a gas are halved, the volume is:

    A. be halved   **C.** double

    B. remain the same   **D.** quadruple   **E.** decrease by a factor of 4

112. According to Gay–Lussac's law, what happens to a gas as temperature increases?

    A. The pressure increases   **C.** The volume increases

    B. The volume decreases   **D.** The pressure decreases

                                     E. none of the above

113. Which one of these molecules can act as a hydrogen bond acceptor but not a donor?

    A. CH_3NH_2 **B.** CH_3CO_2H **C.** H_2O **D.** C_2H_5OH **E.** $CH_3–O–CH_3$

114. A mixture of gases contains 16 g of O_2, 14 g of N_2 and 88 g of CO_2 is collected above water at a temperature of 25 °C. The total pressure is 1 atm and the vapor pressure water at 25 °C is 40 torr. What is the partial pressure exerted by CO_2?

    A. 280 torr   **B.** 360 torr   **C.** 480 torr   **D.** 540 torr   **E.** 560 torr

115. What is the mole fraction of H in a gaseous mixture that consists of 8.00 g of H_2 and 12.00 g of Ne in a 3.50 liter container maintained at 35.20 °C?

    A. 0.130   **B.** 0.430   **C.** 0.670   **D.** 0.870   **E.** 0.910

116. A gas initially filled a 3.0 L container? Heat was then added to the gas which raised its temperature from 100 K to 150 K while increasing its pressure from 3.0 atm to 4.5 atm. What is the new volume of the gas?

    A. 1.4 L   **B.** 3.0 L   **C.** 2.0 L   **D.** 4.5 L   **E.** 6.0 L

117. How will the pressure of a fixed sample of gas change if its volume is halved and Kelvin is quadrupled while the moles of gas remain constant?

    A. decreases by a factor of 8   **C.** decreases by a factor of 4

    B. quadruples   **D.** decreases by a factor of 2

                                     E. increases by a factor of 8

                                      Copyright © 2015 Sterling Test Prep

118. If a piston compresses a gas in a steel cylinder, what happens to the volume in the cylinder?

- **A.** increases and the pressure decreases
- **B.** increases and the pressure increases
- **C.** decreases and the pressure decreases
- **D.** decreases and the pressure increases
- **E.** none of the above

119. What is the predominant intermolecular force that is between two molecules of CH_3CH_2OH?

- **A.** ion–dipole
- **B.** ion–ion
- **C.** dipole–dipole
- **D.** London dispersion forces
- **E.** hydrogen bonding

120. As the pressure is increased on solid CO_2, the melting point is:

- **A.** decreased
- **B.** unchanged
- **C.** increased
- **D.** inversely proportional to the square root of the change
- **E.** unable to be determined without further information

121. The average speed at which a methane molecule effuses at 28.5 °C is 631 m/s. The average speed at which a krypton molecule effuses at this same temperature is:

- **A.** 123 m s^{-1}
- **B.** 276 m s^{-1}
- **C.** 312 m s^{-1}
- **D.** 421 m s^{-1}
- **E.** 633 m s^{-1}

122. A sample of SO_3 gas is decomposed to SO_2 and O_2.

$$2SO_3(g) \rightarrow 2SO_2(g) + O_2(g)$$

If the total pressure of SO_2 and O_2 is 1340 torr, what is the partial pressure of O_2 in torr?

- **A.** 447 torr
- **B.** 1160 torr
- **C.** 1340 torr
- **D.** 893 torr
- **E.** 11.60 torr

123. Which of the following is the definition of standard temperature and pressure?

- **A.** 0 K and 1 atm
- **B.** 298 K and 760 mm Hg
- **C.** 298 K and 1 atm
- **D.** 273 °C and 760 torr
- **E.** 273 K and 760 mm Hg

124. A latex balloon has a volume of 500 mL when filled with gas at a pressure of 820 torr and a temperature of 300 K. The balloon contains how many moles of gas? ($R = 0.08206 \text{ L atm mol}^{-1} \text{ K}^{-1}$)

- **A.** 0.022
- **B.** 0.22
- **C.** 2.2
- **D.** 22
- **E.** 222

125. Avogadro's law, in its alternate form, is very similar in form to:

- **A.** Charles' law and Gay-Lussac's law
- **B.** Boyle's law and Charles' law
- **C.** Boyle's law
- **D.** Boyle's law and Gay-Lussac's law
- **E.** Charles' law

126. If a helium balloon is placed in a cold freezer, what happens to the temperature in the balloon?

    A. increases and the volume increases   **C.** decreases and the volume increases

    B. increases and the volume decreases   **D.** decreases and the volume decreases

                                                E. none of the above

127. Which of the following compounds exhibits primarily dipole-dipole intermolecular forces?

    A. CO_2 **B.** F_2 **C.** $CH_3–O–CH_3$ **D.** CH_3CH_3 **E.** H_2O

128. Boiling occurs when the:

    A. internal pressure of a liquid is less than the sum of all external pressures

    B. vapor pressure of a liquid is greater than the external pressure

    C. internal pressure of a liquid is greater than the atmospheric pressure

    D. internal pressure of a liquid is greater than the external pressure

    E. vapor pressure of a liquid is less than the external pressure

129. Which of the following two variables are present in mathematical statements of Avogadro's law?

    A. P and V   **B.** n and P   **C.** n and T   **D.** V and T   **E.** n and V

130. As an automobile travels the highway, what happens to the temperature of the air inside the tires?

    A. increases and the pressure decreases   **C.** decreases and the pressure decreases

    B. increases and the pressure increases   **D.** decreases and the pressure increases

                                                E. none of the above

131. How many hydrogen bonds does $CH_3–O–CH_2OH$ form with water?

    A. 5   **B.** 3   **C.** 4   **D.** 2   **E.** 6

132. Which statement is NOT true regarding vapor pressure?

    A. Solids have a vapor pressure

    B. The vapor pressure of a pure liquid does not depend on the amount of vapor present

    C. The vapor pressure of a pure liquid does not depend on the amount of liquid present

    D. None of the above

    E. A and B

133. According to the kinetic theory of gases, the average kinetic energy of the gas particles is directly proportional to:

    A. temperature   **C.** volume

    B. molar mass   **D.** pressure   **E.** number of moles of gas

Copyright © 2015 Sterling Test Prep

134. How many moles of gas are present in a 10.0 liter sample at STP?

    A. 224 moles   **C.** 10.0 moles

    B. 0.446 moles   **D.** 2.24 moles   **E.** 22.4 moles

135. A balloon contains 40 grams of He with a pressure of 1000 torr. He was released from the balloon and the new pressure was 900 torr and the volume is half of the original. If the temperature remains the same, how many grams of He is in the balloon?

    A. 18 grams   **B.** 20 grams   **C.** 10 grams   **D.** 40 grams   **E.** 25 grams

136. Which is the strongest form of intermolecular attraction in a water molecule?

    A. ion-dipole   **C.** induced dipole-induced dipole

    B. covalent bonding   **D.** hydrogen bonding

                            E. polar-induced polar

137. An ideal gas differs from a real gas because the molecules of an ideal gas have:

    A. no attraction to each other   **C.** molecular weight equal to zero

    B. no kinetic energy   **D.** appreciable volumes

                            E. none of the above

138. Which of the following describes a substance in the solid physical state?

    A. The substance compresses negligibly   **C.** The substance has a fixed shape

    B. The substance has a fixed volume   **D.** None of the above

                                    E. All of the above

139. For small molecules of comparable molecular weight, which lists the intermolecular forces in increasing order?

    A. dipole-dipole forces < hydrogen bonds < London forces

    B. London forces < hydrogen bonds < dipole-dipole forces

    C. London forces < dipole-dipole forces < hydrogen bonds

    D. hydrogen bonds < dipole-dipole forces < London forces

    E. London forces < dipole-dipole forces = hydrogen bonds

140. Consider the phase diagram for H_2O. The termination of the gas-liquid transition at which distinct or liquid phases do not exist is called the:

    A. critical point   **C.** triple point

    B. end point   **D.** condensation point   **E.** inflection point

141. According to Avogadro's Law, the volume of a gas [ ] as the [ ] increases while [ ] are held constant.

    A. increases… temperature… pressure and number of moles

    B. decreases… pressure… temperature and number of moles

    C. increases… pressure… temperature and number of moles

    D. decreases… number of moles… pressure and temperature

    E. increases… number of moles… pressure and temperature

142. The ideal gas equation can NOT be written as:

   A. $PV = nRT$ **C.** $R = PV / nT$

   B. $P = nRT / V$ **D.** $R = nT / PV$ **E.** none of the above

143. Which of the following describes a substance in the liquid physical state?

   A. The substance has a variable shape   **C.** The substance has a fixed volume

   B. The substance compresses negligibly   **D.** None of the above

                                        E. All of the above

144. Which has only London dispersion forces as the primary attraction between molecules?

   A. CH_3OH **C.** CH_3CH_3

   B. CH_3CH_2OH **D.** H_2S **E.** H_2O

145. Which one does NOT demonstrate colligative properties?

   A. Freezing point   **C.** Vapor pressure

   B. Boiling point   **D.** All of the above demonstrate colligative properties

                                        E. none of the above

146. If container X is occupied by 1.00 mole of O_2 gas while container Y is occupied by 20.0 grams of N_2 gas and both containers are maintained at 0.00 °C and 650 torr then:

   A. container X must have a volume of 22.4 L

   B. the average kinetic energy of the molecules in X is equal to the average kinetic energy of the molecules in Y

   C. container Y must be larger than container X

   D. the average speed of the molecules in container X is greater than that of the molecules in container Y

   E. the number of atoms in container Y is greater than the number of atoms in container X

147. Consider a sample of helium and a sample of neon, both at 30 °C and 1.5 atm. Both samples have a volume of 5.0 liters. Which statement regarding these samples is NOT true?

   A. The density of the neon is greater than the density of the helium

   B. Each sample contains the same number of moles of gas

   C. Each sample weighs the same amount

   D. Each sample contains the same number of atoms of gas

   E. none of the above

148. 10 grams of O_2 are placed in an empty 10 liter container at 28 °C. Compared to an equal mass of H that is placed in an identical container at 28 °C, the pressure of the O_2 is:

   A. less than the pressure of the H   **C.** greater than the pressure of the H

   B. double the pressure of the H   **D.** equal to the pressure of the H

                                        E. equal to the square root of the pressure of the H

149. Which molecules is most likely to show a dipole-dipole interaction?

   A. CH_4 **B.** $H–C{\equiv}C–H$ **C.** SO_2 **D.** CO_2 **E.** none of the above

                                       Copyright © 2015 Sterling Test Prep

150. The value of the ideal gas constant, expressed in units (torr × mL) / mole × K is:

   A. 62.4   **B.** 62,400   **C.** 0.0821   **D.** 1 / 0.0821   **E.** 8.21

151. What is the term for a change of state from a liquid to a gas?

   A. vaporizing   **C.** freezing

   B. melting   **D.** condensing   **E.** none of the above

152. Which of the following pairs of compounds contain the same intermolecular forces?

   A. H_2S and CH_4 **C.** CH_3CH_3 and H_2O

   B. NH_3 and CH_4 **D.** CH_3CH_2OH and H_2O

                                     E. CCl_4 and CH_3OH

153. When nonvolatile solute molecules are added to a solution, the vapor pressure of the solution:

   A. stays the same   **C.** decreases

   B. increases   **D.** is unable to be determined without further information

                            E. directly proportional to the square root of the amount added

154. For the following reaction, if 11.2 L of nitrogen are reacted to form NH_3 at STP, how many liters of hydrogen are required to completely consume all of the nitrogen?

      $N_2 + 3H_2 \rightarrow 2NH_2$

   A. 7.4 L   **B.** 6.1 L   **C.** 16.8 L   **D.** 11.2 L   **E.** 33.6 L

155. What volume of H_2 gas at 780 mm Hg and 23 °C is required to produce 10.6 L of NH_3 gas at the same temperature and pressure using the following chemical reaction?

      $N_2 (g) + 3H_2 (g) \rightarrow 2NH_3 (g)$

   A. 20.0 L   **B.** 15.0 L   **C.** 15.9 L   **D.** 13.0 L   **E.** 10.6 L

156. What is the term for a direct change of state from a solid to a gas?

   A. sublimation   **C.** condensation

   B. vaporization   **D.** deposition   **E.** none of the above

157. What is the molarity of a solution that contains 36 mEq Ca^{2+} per liter?

   A. 0.018 M   **B.** 0.036 M   **C.** 1.8 M   **D.** 3.6 M   **E.** 0.36 M

158. What effect does adding solute to a pure solvent have on the boiling and freezing points, respectively?

   A. decreases, decreases   **C.** increases, decreases

   B. decreases, increases   **D.** increases, increases   **E.** no relationship

159. The van der Waals equation of state for a real gas is

$$\left[P + \frac{n^2 a}{V^2}\right](V - nb) = nRT$$

The van der Waals constant, *a*, represents a correction for a:

A. negative deviation in the measured value of P from that for an ideal gas due to the attractive forces between the molecules of a real gas

B. positive deviation in the measured value of P from that for an ideal gas due to the attractive forces between the molecules of a real gas

C. negative deviation in the measured value of P from that for an ideal gas due to the finite volume of space occupied by molecules of a real gas

D. positive deviation in the measured value of P from that for an ideal gas due to the finite volume of space occupied by molecules of a real gas

E. positive deviation in the measured value of P from that for an ideal gas due to the finite mass of the molecules of a real gas

160. Vessels X and Y each contain 1.00 L of a gas at STP, but vessel X contains oxygen while container Y contains nitrogen. Assuming the gases behave ideally, the gases has the same:

I. number of molecules   II. density   III. kinetic energy

A. II only   C. I and III only

B. III only   D. I, II, and III   E. I only

161. Which molecules is most likely to show a dipole-dipole interaction?

I. H–C≡C–H   II. CH_4 III. CH_3SH IV. $CH_3 CH_2OH$

A. I only   B. II only   C. III only   D. III and IV only   E. I and IV only

162. Propane burners are used for cooking by campers. What volume of H_2O gas is produced by the complete combustion of 1.8 L of propane (C_3H_8) gas? Gas volumes are measured at the same temperature and pressure.

$$C_3H_8(g) + 5O_2(g) \rightarrow 3CO_2(g) + 4H_2O(g)$$

A. 0.52 L   B. 7.2 L   C. 14 L   D. 1.8 L   E. 5.2 L

163. What is the term for a change of state from a gas to a liquid?

A. vaporizing   C. freezing

B. melting   D. condensing   E. none of the above

164. What is the molarity of the solution obtained by diluting 125 mL of 2.50 M NaOH to 575 mL?

A. 0.272 M   B. 0.543 M   C. 1.84 M   D. 1.15 M   E. 0.115 M

165. What is the boiling point when 90 g of glucose ($C_6H_{12}O_6$) are added to 500 g of H_2O?

K= 0.52

A. 267.80 K   B. 273.54 K   C. 278.20 K   D. 373.52 K   E. 393.28 K

 Copyright © 2015 Sterling Test Prep

166. Currently the CO_2 content of the Earth's atmosphere is approximately:

    A. 40.5 %   **B.** 11 %   **C.** 4.21 %   **D.** 0.038 %   **E.** 0.011 %

167. A container is labeled "Ar, 5.0 moles" but has no pressure gauge. By measuring the temperature and determining the volume of the container, the chemist uses the ideal gas law to estimate the pressure inside the container. Unfortunately, the handwriting on the container is illegible, and the container actually contains 5.0 moles of He, not Ar. How does this affect the pressure inside of the container?

    A. The estimate is correct but only because both gases are monatomic

    B. The estimate is too high

    C. The estimate is slightly too low

    D. The estimate is significantly too low because of the large difference in molecular mass

    E. The estimate is correct because the identity of the gas is irrelevant

168. Which is most likely to have the weakest induced dipole-induced dipole interaction?

    A. I_2 **B.** F_2 **C.** Br_2 **D.** Cl_2 **E.** All have the same interactions

169. How much water must be added when 125 mL of a 2.00 M solution of HCl is diluted to a final concentration of 0.400 M?

    A. 150 mL   **B.** 850 mL   **C.** 625 mL   **D.** 750 mL   **E.** 500 mL

170. The freezing point changes by 10 K when an unknown amount of toluene is added to 100 g of benzene. Find the number of moles of toluene added from the given data.

    $K_{benzene} = 5.0$

    $K_{toluene} = 8.4$

    A. 0.14   **B.** 0.20   **C.** 0.23   **D.** 0.27   **E.** 0.31

171. A laser contains 0.40 mole Ar and 0.60 mole F_2 whereby the total pressure inside the laser cavity is 1.10 atm. What is the partial pressure of Ar inside the laser?

    A. 0.40 atm   **C.** 0.60 atm

    B. 10.0 atm   **D.** 1.0 atm   **E.** 0.44 atm

172. Which of the following atoms could interact through a hydrogen bond?

    A. The hydrogen of an amine and the oxygen of an alcohol

    B. The hydrogen on an aromatic ring and the oxygen of carbon dioxide

    C. The oxygen of a ketone and the hydrogen of an aldehyde

    D. The oxygen of methanol and a hydrogen on the methyl carbon of methanol

    E. None of the above

173. The ozone layer is located in which region of the atmosphere?

    A. troposhere   **C.** mesosphere

    B. stratosphere   **D.** thermosphere   **E.** ecosphere

174. Which of the following statements concerning these gases is true if three 2.0 L flasks are filled with H_2, O_2 and He, respectively, at STP?

    A. There are twice as many He atoms as H_2 or O_2 molecules

    B. There are four as many H_2 or O_2 molecules as He atoms

    C. Each flask contains the same number of atoms

    D. There are twice as many H_2 or O_2 molecules as He atoms

    E. The number of H_2 or O_2 molecules is the same as the number of He atoms

175. What is the molarity of a solution formed by dissolving 45 g NaOH in water to give a final volume of 250 mL?

    A. 0.0045 M   **B.** 0.18 M   **C.** 4.5 M   **D.** 9.0 M   **E.** 0.45 M

176. When 25 g of compound X is dissolved in 1 kg of camphor, the freezing point of the camphor falls 2.0 K. What is the approximate molecular weight of compound X?

        $K_{camphor} = 40$

    A. 50 g/mol   **C.** 5,000 g/mol

    B. 500 g/mol   **D.** 5,500 g/mol   **E.** 5,750 g/mol

177. Which of the following best explains the hydrogen bonding that occurs in water?

    A. The structure of liquid water is best described as flickering clusters of H-bonds due to the relative short duration of individual H-bonds

    B. Each water molecule is capable of forming 8 H-bonds: 2 from each lone pair of electrons and 2 from each hydrogen

    C. The average number of H-bonds formed by one water molecule is the same in liquid and solid water, the only difference is the duration of the H-bond

    D. The number of H-bonds formed by one water molecule is greater in liquid water than in solid water

    E. None of the above

178. What volume is occupied by 6.21×10^{24} molecules of CO at STP?

    A. 231 L   **B.** 44.3 L   **C.** 22.4 L   **D.** 106 L   **E.** 2.24 L

179. According to the kinetic theory, what happens to the kinetic energy of gaseous molecules when the temperature of a gas decreases?

    A. increase as does velocity   **C.** increases and velocity decreases

    B. remains constant as does velocity   **D.** decreases and velocity increases

                                           E. decreases as does velocity

180. Calculate the volume (in ml) of a 2.75 M solution that must be used to make 1.25 L of a 0.150 M solution.

    A. 0.0682 mL   **B.** 33.0 mL   **C.** 0.0330 mL   **D.** 68.2 mL   **E.** 0.682 mL

Copyright © 2015 Sterling Test Prep

181. Calculate the number of O_2 molecules if a 15.0 L cylinder was filled with O_2 gas at STP.

    A. 443 molecules

    B. 6.59×10^{24} molecules

    C. 4.03×10^{23} molecules

    D. 2.77×10^{22} molecules

    E. 4,430 molecules

182. Given the structure of glucose, which explains the hydrogen bonding between glucose and water?

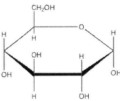

    A. Due to the cyclic structure of glucose, H-bonding with water does not occur

    B. Each glucose molecule could H-bond with as many as 17 water molecules

    C. H-bonds will form with water always being the H-bond donor

    D. H-bonds will form with glucose always being the H-bond donor

    E. None of the above

183. The conditions known as STP are:

    A. 1 mm Hg and 273 K

    B. 1 atm and 273 °C

    C. 760 mm Hg and 273 °C

    D. 760 atm and 273 K

    E. 1 atm and 273 K

184. For a gas to be classified as a *greenhouse gas*, it must:

    A. transmit infrared light and absorb visible light

    B. be combustible

    C. transmit visible light and absorb infrared radiation

    D. be radioactive

    E. be a product of combustion

185. Which of the following would have the highest boiling point?

    A. F_2 **B.** Br_2 **C.** Cl_2 **D.** I_2

    E. not enough information given

186. A sample of N_2 gas occupies a volume of 180 mL at STP. What volume will it occupy at 640 mmHg and 295 K?

    A. 1.07 L   **B.** 0.690 L   **C.** 0.231 L   **D.** 1.45 L   **E.** 6.90 L

187. What volume of 8.25 M NaOH solution must be diluted to prepare 2.40 L of 0.500 M NaOH solution?

    A. 438 mL   **B.** 145 mL   **C.** 39.6 L   **D.** 0.356 L   **E.** 35.4 L

188. A properly designed Torricelli mercury barometer should be at least how tall?

    A. 100 in   **B.** 250 mm   **C.** 76 mm   **D.** 500 mm   **E.** 800 mm

189. Which gas has the greatest density at STP:

    A. CO_2 **B.** O_2 **C.** N_2 **D.** NO   **E.** more than one

190. A container contains only N_2, CO_2, O_2 and water vapor. If, at STP, the partial pressure of O_2 is 300 torr, CO_2 20 torr, and water vapor 8 torr, what is the partial pressure of N_2?

    A. 165 torr   **B.** 432 torr   **C.** 760 torr   **D.** 330 torr   **E.** 864 torr

191. What is the mass of one mole of a gas that has a density of 1.34 g/L at STP?

    A. 44.0 g   **B.** 43.9 g   **C.** 28.0 g   **D.** 30.0 g   **E.** 4.39 g

192. Explain why chlorine, Cl_2, is a gas at room temperature while bromine, Br_2, is a liquid.

    A. Bromine ions are held together by ionic bonds

    B. Chlorine molecules are smaller and therefore pack tighter in their physical orientation

    C. Bromine atoms are larger which causes the formation of a stronger induced dipole-induced dipole attraction

    D. Chlorine atoms are larger which causes the formation of a stronger induced dipole-induced dipole attraction

    E. Bromine molecules are smaller and therefore pack tighter in their physical orientation

Copyright © 2015 Sterling Test Prep

Chapter 4. SOLUTION CHEMISTRY

1. Which of the following solid compounds is insoluble in water?

    A. $BaSO_4$ **C.** $PbCl_2$

    B. Hg_2I_2 **D.** all of the above   **E.** none of the above

2. The sucrose that dissolves in H_2O is the [   ], the H_2O is the [   ], and the sweetened H_2O is the [   ], respectively.

    A. solution, solute, solvent   **C.** solute, solvent, solution

    B. solute, solution, solvent   **D.** solvent, solution, solute

                                   E. solution, solvent, solute

3. The equation for the reaction $Pb(NO_3)_2\,(aq) + NaCl\,(aq) \rightarrow PbCl_2\,(s) + NaNO_3\,(aq)$ can be written as an ionic equation. In the ionic equation, the spectator ions are:

    A. Pb^{2+} and NO_3^- **C.** Na^+ and Cl^-

    B. Pb^{2+} and Cl^- **D.** Na^+ and Pb^{2+} **E.** Na^+ and NO_3^-

4. All of the following statements describing solutions are true, EXCEPT:

    A. solutions are colorless

    B. the particles in a solution are atomic or molecular   **D.** solutions are homogeneous

    C. making a solution involves a physical change in size   **E.** solutions are transparent

5. Which of the following is possible?

    I. A gaseous solution consisting of a gaseous solute and a gaseous solvent

    II. A solid solution consisting of a solid solute and a solid solvent

    III. A liquid solution resulting from a gaseous solute and a liquid solvent

    A. I only   **B.** II only   **C.** III only   **D.** I and II only   **E.** I, II and III

6. A crystal of solid NaCl is placed into an aqueous NaCl solution. No precipitate forms in the bottom of the container and the final solution is:

    I. slightly saturated   II. unsaturated   III. supersaturated

    A. I only   **C.** III only

    B. I and II only   **D.** I and III only   **E.** I, II and III

7. Which of the following solid compounds is insoluble in water?

    A. Li_2CO_3 **C.** $AgC_2H_3O_2$

    B. $Cu(NO_3)_2$ **D.** all of the above   **E.** none of the above

8. What is the molar concentration of a solution containing 0.5 mol of solute in a 50 cm^3 of solution?

    A. 0.1 M   **B.** 1.5 M   **C.** 3 M   **D.** 10 M   **E.** 1 M

Copyright © 2015 Sterling Test Prep

9. Which of the following is/are solutions?

        I. NaCl in water

        II. An alloy of 2% carbon and 98% water

        III. Water vapor in nitrogen gas

   A. I only   **B.** II only   **C.** II and III only   **D.** I, II and III   **E.** I and II only

10. In a solution made from one teaspoon of sugar and one liter of water, which is the solute?

   A. the teaspoon   **C.** sugar

   B. water   **D.** both sugar and water   **E.** none of the above

11. Which of the following solid compounds is insoluble in water?

   A. $BaSO_4$ **C.** $(NH_4)_2CO_3$

   B. Na_2S **D.** K_2CrO_4 **E.** $Sr(OH)_2$

12. A saturated solution:

   A. contains dissolved solute in equilibrium with undissolved solid

   B. will rapidly precipitate if a seed crystal is added   **D.** contains only electrolytes

   C. contains as much solvent as it can hold   **E.** contains only ions

13. The equation for the reaction $AgNO_3\,(aq) + K_2CrO_4\,(aq) \rightarrow Ag_2CrO_4\,(s) + KNO_3\,(aq)$ can be written as an ionic equation, whereby the spectator ions are:

   A. K^+ and CrO_4^{2-} **C.** Ag^+ and K^+

   B. CrO_4^{2-} and NO_3^- **D.** Ag^+ and CrO_4^{2-} **E.** K^+ and NO_3^-

14. Which homogeneous mixture is opaque and has particles large enough to be filtered?

   A. colloid   **C.** solution

   B. suspension   **D.** both colloids and suspensions

                      E. none of the above

15. Which are characteristics of an ideally dilute solution?

        I. Solute molecules do not interact with each other

        II. Solvent molecules do not interact with each other

        III. The mole fraction of the solvent approaches 1

   A. I only   **C.** II and III only

   B. II only   **D.** I, II and III   **E.** I and II only

16. A solution in which the rate of crystallization is equal to the rate of dissolution is:

   A. saturated   **C.** dilute

   B. supersaturated   **D.** unsaturated   **E.** cannot be determined

17. Which of the following symbolizes a precipitate in a chemical equation?

   A. *(g)*   **B.** *(aq)*   **C.** *(s)*   **D.** *(l)*   **E.** none of the above

                                                   Copyright © 2015 Sterling Test Prep

18. What is the molality of a solution that contains 2 moles of glycerin, $C_3H_5(OH)_3$, dissolved in 1000g of water?

    A. 0.2 m   **B.** 0.5 m   **C.** 1 m   **D.** 2 m   **E.** 4 m

19. When 1 ml of a 4M solution of HCl is diluted to 15 ml, the new concentration of the HCl is:

    A. 0.140 M   **B.** 0.440 M   **C.** 0.325 M   **D.** 0.415 M   **E.** 0.270 M

20. Which is/are characteristic(s) of an ideal solution?

    I. Solute molecules do not interact
    II. Solvent molecules do not interact
    III. Solvent-solute interactions are similar to solute-solute and solvent-solvent interactions

    A. I only   **B.** III only   **C.** II only   **D.** I, II and III   **E.** I and II only

21. In a solution of 77 percent nitrogen and 23 percent oxygen, which is the solvent?

    A. nitrogen   **C.** both

    B. oxygen   **D.** neither   **E.** gases cannot form solutions

22. A supersaturated solution:

    A. contains dissolved solute in equilibrium with undissolved solid

    B. will rapidly precipitate if a seed crystal is added   **D.** contains no double bonds

    C. contains as much solvent as it can hold   **E.** contains only electrolytes

23. Which statement is correct for pure H_2O?

    A. Pure H_2O contains no ions

    B. Pure H_2O is an electrolyte

    C. Pure H_2O contains equal $[^-OH]$ and $[H_3O^+]$

    D. Pure H_2O contains greater $[^-OH]$ than $[H_3O^+]$

    E. Pure H_2O contains greater $[H_3O^+]$ than $[^-OH]$

24. Which compound produces three ions per formula unit by dissociation when dissolved in water?

    A. aluminum sulfate   **C.** nickel sulfate

    B. ammonium bromate   **D.** sodium nitrate   **E.** calcium perchlorate

25. Which compound is most likely to be more soluble in benzene (a nonpolar solvent) than in water?

    A. SO_2 **B.** CO_2 **C.** silver chloride   **D.** H_2S **E.** CH_2Cl_2

26. Which of the following statements describes a saturated solution?

    A. Solution where the solvent cannot dissolve any more solute

    B. Carbonated beverage with bubbles   **D.** All of the above

    C. Solution of salt water with salt at the bottom   **E.** None of the above

27. What is the molarity of a solution formed by dissolving 25.0 g of NaCl in water to make 625 mL of solution?

    A. 0.495 M   **B.** 0.684 M   **C.** 0.328 M   **D.** 0.535 M   **E.** 0.125 M

28. Which statement comparing solutions with pure solvents is NOT correct?

    A. A solution containing a non-volatile solute has a lower boiling point than a pure solvent

    B. A solution will have a greater mass than an equal volume of a pure solvent if the solute has a molar mass greater than the solvent

    C. A solution containing a non-volatile solute has a lower vapor pressure than a pure solvent

    D. A solution containing a non-volatile solute has a lower freezing point than a pure solvent

    E. None of the above

29. Which of the following is the most soluble in benzene (C_6H_6)?

    A. glucose ($C_6H_{12}O_6$)   **C.** octane (C_8H_{18})

    B. sodium benzoate   **D.** hydrobromic acid   **E.** CH_2Cl_2

30. The term miscible describes which type of solution?

    A. solid/solid   **C.** liquid/solid

    B. liquid/gas   **D.** liquid/liquid   **E.** solid/gas

31. What is the predicted product from the burning of carbon in charcoal?

    A. CO_3 **C.** C_2O

    B. CO_2 **D.** CO   **E.** The product depends on experimental conditions

32. Approximate the freezing point of an aqueous solution containing 25 g of a compound with a molecular weight of 100 g/mol dissolved in 50 g of water. (Note: assume the compound does not dissociate in solution. Molal freezing point constant of water = -1.86 °C/m).

    A. -4.5 °C   **B.** -14 °C   **C.** -9 °C   **D.** -18 °C   **E.** -21 °C

33. Which of the following can serve as the solvent in a solution?

    A. liquid   **C.** solid

    B. gas   **D.** liquid and gas   **E.** All of the above

34. Which is most likely soluble in NH_3?

    A. CO_2 **B.** SO_2 **C.** CCl_4 **D.** N_2 **E.** H_2

35. Water and methanol are two liquids that dissolve in each other. When the two are mixed they form one layer because the liquids are:

    A. unsaturated   **B.** saturated   **C.** miscible   **D.** immiscible   **E.** supersaturated

36. What is the predicted product from the burning of sulfur in low-grade coal?

    A. SO_2 **C.** SO

    B. SO_3 **D.** S_2O **E.** The product depends on experimental conditions

      Copyright © 2015 Sterling Test Prep

37. What volume of 10 M H_2SO_4 is needed to prepare 600 ml of 0.5M H_2SO_4?

   A. 5 ml   **B.** 10 ml   **C.** 30 ml   **D.** 15 ml   **E.** 60 ml

38. A solute is a:

   A. substance that dissolves into a solvent
   B. substance containing a solid, liquid or gas
   C. solid substance that does not dissolve into water
   D. solid substance that does not dissolve at a given temperature
   E. liquid that does not dissolve into another liquid

39. Salts are more soluble in H_2O than in benzene because:

   A. benzene is aromatic and therefore very stable
   B. the dipole moment of H_2O compensates for the loss of ionic bonding when salt dissolves
   C. the strong intermolecular attractions in benzene must be disrupted to dissolve a salt in benzene
   D. the molecular mass of H_2O is similar to the atomic mass of most ions
   E. the dipole moment of H_2O compensates for the increased ionic bonding when salt dissolves

40. Which of the following interparticle attractions play a part in the formation of a solution?

     I. solute-solute   II. solvent-solute   III. solvent-solvent

   A. I only   **C.** I and II only
   B. III only   **D.** I, II and III   **E.** II and III only

41. What is the product from the reaction of N_2 and O_2 gases in a combustion engine?

   A. N_2O_3 **B.** NO_2 **C.** N_2O **D.** NO   **E.** all of the above

42. Which has the highest boiling point?

   A. 0.1 M $Al(NO_3)_3$ **C.** 0.1 M glucose ($C_6H_{12}O_6$)
   B. 0.1 M $MgCl_2$ **D.** 0.1 M Na_2SO_4 **E.** pure H_2O

43. Octane is less soluble in H_2O than in benzene because:

   A. bonds between benzene and octane are much stronger than the bonds between H_2O and octane
   B. octane cannot dissociate in the presence of H_2O
   C. bonds between H_2O and octane are weaker than the bonds between H_2O molecules
   D. octane and benzene have similar molecular weights
   E. H_2O dissociates in the presence of octane

44. Which of the following is true of a colloid?

   A. Dispersed particles separate in a centrifuge
   B. Dispersed particles pass through filter paper   **D.** All of the above
   C. Dispersed particles are about 1-100 nm in diameter   **E.** None of the above

45. What is the volume of a solution that contains 3.12 moles of NaCl if the concentration is 6.67 M NaCl?

   A. 0.936 L   **B.** 0.468 L   **C.** 20.8 L   **D.** 2.14 L   **E.** 0.234 L

46. Which of the following describes the term concentration?

   A. It is the amount of solute in a given amount of solution

   B. The given amount of solution in a given container

   C. The given amount of solvent per amount of solute

   D. The amount of solvent in a given amount of solution

   E. None of the above

47. Methane (CH_4) can be used as automobile fuel to reduce pollution. What is the coefficient of oxygen in the balanced equation for the reaction?

$$\underline{\quad}CH_4\,(g) + \underline{\quad}O_2\,(g) \xrightarrow{\text{Spark}} \underline{\quad}CO_2\,(g) + \underline{\quad}H_2O\,(g)$$

   A. 1   **B.** 2   **C.** 3   **D.** 4   **E.** none of the above

48. Determine the boiling point of a 2 molal aqueous solution of NaCl.

Note: boiling point constant of water = 0.51 °C/m

   A. 98.5 °C   **B.** 99.8 °C   **C.** 101 °C   **D.** 102 °C   **E.** 105 °C

49. Of the following, which can serve as the solute in a solution?

   A. gas   **C.** liquid

   B. solid   **D.** gas and liquid only   **E.** all of the above

50. Hexane is significantly soluble in octane because:

   A. entropy increases for the two substances as the dominant factor in the ΔG when mixed

   B. hexane hydrogen bonds with octane

   C. intermolecular bonds between hexane-octane are much stronger than either hexane-hexane or octane-octane molecular bonds

   D. ΔH for hexane-octane is greater than hexane-H_2O

   E. hexane and octane have similar molecular weights

51. The principle *like dissolve like* is not applicable for predicting solubility when the solute is a/an:

   A. nonpolar liquid   **C.** nonpolar gas

   B. polar gas   **D.** ionic compound   **E.** covalent compound

52. Ethane (C_2H_6) gives carbon dioxide and water when burning. What is the coefficient of oxygen in the balanced equation for the reaction?

$$\underline{\quad}C_2H_6\,(g) + \underline{\quad}O_2\,(g) \xrightarrow{\text{Spark}} \underline{\quad}CO_2\,(g) + \underline{\quad}H_2O\,(g)$$

   A. 5   **B.** 7   **C.** 10   **D.** 14   **E.** none of the above

53. A 4M solution of H_3A is completely dissociated in water. How many equivalents of H^+ are found in 1/3 liter?

   A. ¼   **B.** 1   **C.** 1.5   **D.** 3   **E.** 4

Copyright © 2015 Sterling Test Prep

54. Which of the following describes a saturated solution?

   A. When the ratio of solute to solvent is small
   B. When it contains less solute than it can hold at 25 °C
   C. When it contains as much solute as it can hold at a given temperature
   D. When it contains 1 g of solute in 100 mL of water
   E. When it is equivalent to a supersaturated solution

55. 15 grams of an unknown substance is dissolved in 60 grams of water. When the solution is transferred to another container, it weighs 78 grams. Which of the following is a possible explanation?

   A. The solution reacted with the second container that formed a precipitate
   B. Some of the solution remained in the first container
   C. The reaction was endothermic, which increased the average molecular speed
   D. The solution reacted with the first container, causing some byproducts to be transferred with the solution
   E. The reaction was exothermic, which increased the average molecular speed

56. Which of the following solutions is the most dilute?

   A. One liter of H_2O with 1 gram of sugar
   B. One liter of H_2O with 2 grams of sugar
   C. One liter of H_2O with 5 grams of sugar
   D. One liter of H_2O with 10 grams of sugar
   E. All have the same volume

57. Which type of compound is likely to dissolve in H_2O?

   I. One with hydrogen bonds
   II. Highly polar compound
   III. Salt

   A. I only   **B.** II only   **C.** III only   **D.** I and III only   **E.** I, II and III

58. Propane (C_3H_8) is flammable and used in rural areas where natural gas is not available. What is the coefficient of oxygen in the balanced equation for the combustion of propane?

$$\underline{\quad}C_3H_8\,(g) + \underline{\quad}O_2\,(g) \xrightarrow{\text{Spark}} \underline{\quad}CO_2\,(g) + \underline{\quad}H_2O\,(g)$$

   A. 1   **B.** 7   **C.** 5   **D.** 10   **E.** none of the above

59. In a mixture of 5 mL water, 10 mL alcohol, and 50 mL acetone, the solvent(s) is/are:

   A. acetone
   B. water
   C. alcohol
   D. acetone and alcohol
   E. alcohol and water

60. Which of the following describes a colloid?

   A. A supersaturated solution of potassium chloride
   B. Carbon dioxide molecules dissolved in a soft drink
   C. Hemoglobin molecules in cytosol
   D. Tiny sand particles dispersed in water which then settle as a precipitate
   E. Food coloring dispersed in water that change the color of the liquid

61. Which response includes all of the following compounds that are soluble in water, and no others?

    I. LiF   II. $PbCl_2$ III. NH_4CH_3COO IV. FeS

    V. $Mn(OH)_2$ VI. $Cr(NO_3)_3$ VII. $CuCO_3$ VIII. $Ni_3(PO_4)_2$

    A. III, IV, V and VIII   **C.** I, III and VI

    B. I, II, V and VIII   **D.** II, IV, V and VII   **E.** I, III and VII

62. Butane (C_4H_{10}) is flammable and used in butane lighters. What is the coefficient of oxygen in the balanced equation for the combustion of butane?

$$\_\_C_4H_{10}\,(g) + \_\_O_2\,(g) \xrightarrow{\text{Spark}} \_\_CO_2\,(g) + \_\_H_2O\,(g)$$

    A. 9   **B.** 13   **C.** 18   **D.** 26   **E.** none of the above

63. Water boils at a lower temperature at high altitudes because:

    A. vapor pressure of water is increased at high altitudes

    B. temperature is higher than at low altitudes

    C. more energy is available to break liquid bonds at high altitudes

    D. atmospheric pressure is lower at high altitudes

    E. atmospheric pressure is lower at low altitudes

64. Which of the following aqueous solutions is a poor conductor of electricity?

    I. sucrose, $C_{12}H_{22}O_{11}$

    II. barium nitrate, $Ba(NO_3)_2$

    III. ethylene glycol, $HOCH_2CH_2OH$

    IV. calcium bromide, $CaBr_2$

    V. ammonium chloride, NH_4Cl

    A. I and III only   **C.** I and E only

    B. I, IV and V only   **D.** III, IV and V only   **E.** II, IV and V only

65. Which of the following methods does NOT extract colloidal particles?

    A. distillation   **C.** evaporation

    B. extraction   **D.** dialysis   **E.** simple filtration

66. Which of the following solutions is the most concentrated?

    A. one liter of water with 1 gram of sugar   **C.** one liter of water with 5 grams of sugar

    B. one liter of water with 2 grams of sugar   **D.** one liter of water with 10 grams of sugar

                                                      E. They all have the same volume

67. Octane (C_8H_{18}) is a major component in gasoline. What is the coefficient of oxygen in the balanced equation for the combustion of octane?

$$\_\_C_8H_{18}\,(g) + \_\_O_2\,(g) \xrightarrow{\text{Spark}} \_\_CO_2\,(g) + \_\_H_2O\,(g)$$

    A. 17   **B.** 25   **C.** 34   **D.** 50   **E.** none of the above

Copyright © 2015 Sterling Test Prep

68. Which of the following is the solvent for a homogenous mixture consisting of 12% ethanol, 28% methanol and 60% water?

   A. Ethanol   **C.** Methanol

   B. Water   **D.** Ethanol and methanol   **E.** Octane

69. Which solution has the greatest osmolarity?

   A. 0.14 M KF   **C.** 0.6 M NaCl

   B. 0.2 M $CaBr_2$ **D.** 0.35 M $AlCl_3$ **E.** 0.10 M KNO_3

70. Which one of the following compounds is NOT soluble in water?

   A. NH_4F **B.** $FeCl_3$ **C.** NaOH   **D.** CH_3OH **E.** CuS

71. Ethanol (C_2H_5OH) is made from fermenting grain and is blended with gasoline as a fuel additive. If combustion of ethanol produces carbon dioxide and water, what is the coefficient of oxygen in the balanced equation?

$$\text{Spark}$$
$$\_\_C_2H_5OH\,(g) + \_\_O_2\,(g) \;\rightarrow\; \_\_CO_2\,(g) + \_\_H_2O\,(g)$$

   A. 1   **B.** 2   **C.** 3   **D.** 6   **E.** none of the above

72. How much NaCl (*s*) (MW = 58) is required to prepare 100 ml of a 4 M solution?

   A. 23.2 g   **B.** 14.5 g   **C.** 232 g   **D.** 145 g   **E.** 15.4 g

73. Which is true when a beam of light shines through a colloid solution?

   A. The path of the light beam is *invisible* in the colloid, but *visible* in the solution
   B. The path of the light beam is *visible* in the colloid, but *invisible* in the solution
   C. The path of the light beam is *invisible* in both the colloid and the solution
   D. The path of the light beam is *visible* in both the colloid and the solution
   E. Depends on the temperature of the experiment

74. Which of the following solutions is the most dilute?

   A. 0.1 liter of H_2O with 1 gram of sugar   **C.** 0.5 liter of H_2O with 5 grams of sugar
   B. 0.2 liter of H_2O with 2 grams of sugar   **D.** 1 liter of H_2O with 10 grams of sugar
                                                E. They all have the same concentration

75. Which of the following is NOT soluble in H_2O?

   A. iron (III) hydroxide   **C.** potassium sulfide
   B. iron (III) nitrate   **D.** ammonium sulfate   **E.** sodium chloride

76. Methanol (CH_3OH) is derived from natural gas and is blended with gasoline. If the combustion of methanol produces carbon dioxide and water, what is the coefficient of oxygen in the balanced equation?

$$\text{Spark}$$
$$\_\_CH_3OH\,(g) + \_\_O_2\,(g) \rightarrow \_\_CO_2\,(g) + \_\_H_2O\,(g)$$

   A. 1   **B.** 2   **C.** 3   **D.** 6   **E.** none of the above

---

77. Calculate the solubility product of AgCl if the solubility of AgCl in H_2O is 1.3×10^{-4} mol/L?

A. 1.3×10^{-4} **C.** 2.6×10^{-4}

B. 1.3×10^{-2} **D.** 3.9×10^{-5} **E.** 1.7×10^{-8}

78. Which statement best explains the meaning of the phrase "like dissolves like"?

A. Only true solutions are formed when water dissolves a nonpolar solute

B. Only true solutions are formed when water dissolves a polar solute

C. A solvent will easily dissolve a solute of similar mass

D. A solvent and solute with similar intermolecular forces will readily form a solution

E. None of these statements is correct

79. What is the term for a homogeneous mixture in which the dispersed particles range from 1 to 100 nm in diameter?

A. colloid   **C.** supersaturated solution

B. true solution   **D.** saturated solution   **E.** none of the above

80. What causes colloid particles to settle out during coagulation?

A. Particles bind to the solution and settle out

B. Particles dissociate and settle out

C. Particles bind together and settle due to gravity

D. Particles dissolve and settle due to gravity

E. Particles break apart and settle due to gravity

81. What statement best describes a mole?

A. The amount of molecules or atoms in 1 gram of something

B. A very large number that chemists use to count atoms or molecules

C. A little furry mammal that lives in the ground

D. A very small number that chemists use to count atoms or molecules

E. None of the above

82. Which percentage concentration units are often used by chemists?

A. % (v/m)   **B.** % (v/v)   **C.** % (m/v)   **D.** % (m/m)   **E.** % (v/d)

83. What is the term for a region in a molecule that has partial negative and partial positive charge resulting from a polar bond?

A. net dipole   **C.** delta minus

B. dipole   **D.** delta plus   **E.** none of the above

84. Which is true if the ion concentration product of a solution of AgCl is less than the K_{sp}?

A. Precipitation occurs   **C.** Precipitation does not occur

B. The ions are insoluble in water   **D.** A and B only   **E.** B and C only

 Copyright © 2015 Sterling Test Prep

85. Which compound produces four ions per formula unit by dissociation when dissolved in water?

    A. Li_3PO_4 **B.** $Ca(NO_3)_2$ **C.** Hg_2Cl_2 **D.** $(NH_4)_2SO_4$ **E.** $(NH_4)_4Fe(CN)_6$

86. What principle states that the solubility of a gas in a liquid is proportional to the partial pressure of the gas above the liquid?

    A. solubility principle   **C.** colloid principle

    B. Tyndall effect   **D.** Henry's law   **E.** none of the above

87. Which of the following represents the chlorite ion?

    A. ClO_4^- **B.** ClO_3^- **C.** ClO^- **D.** Cl_2O^- **E.** ClO_2^-

88. If the concentration of a KCl solution is 16.0% (m/v), the mass of KCl in 26.0 mL of solution is:

    A. 4.16 grams   **C.** 2.08 grams

    B. 8.32 grams   **D.** 16.0 grams   **E.** 8.0 grams

89. What is molarity?

    A. number of liters of solute per mole of solution

    B. number of moles of solute per liter of solvent

    C. number of grams of solute per liter of solution

    D. number of moles of solute per liter of solution

    E. number of grams of solute per mole of solution

90. What is the term that refers to liquids that do not dissolve in one another and separate into two layers?

    A. soluble   **C.** insoluble

    B. miscible   **D.** immiscible   **E.** none of the above

91. The $[Pb^{2+}]$ of a solution is 1.25×10^{-3} M. Calculate the $[SO_4^{2-}]$ that must be exceeded before $PbSO_4$ can precipitate. (The solubility product of $PbSO_4$ at 25 °C is 1.6×10^{-8})

    A. 1.3×10^{-4} M   **C.** 2.6×10^{-4} M

    B. 5.2×10^{-6} M   **D.** 1.6×10^{-8} M   **E.** 1.3×10^{-5} M

92. Which of the following statements best describes the phrase "like dissolves like"?

    A. A solvent dissolves a solute that has a similar mass

    B. Only homogenous solutions are formed when water dissolves a nonpolar solute

    C. A solvent and a solute with similar intermolecular forces make a solution

    D. Only true solutions are formed when hydrophobic solvents dissolve polar solutes

    E. A solvent and a solute with different intermolecular forces make a solution

93. Calculate the molarity of a solution prepared by dissolving 15.0 g of NH_3 in 250 g of water with a final density of 0.974 g/mL.

    A. 60.0 M   **B.** 3.42 M   **C.** 0.00353 M   **D.** 0.882 M   **E.** 6.80 M

94. What is the term for the general principle that solubility is greatest when the polarity of the solute and solvent are similar?

    A. solute rule   **C.** like dissolves like rule

    B. solvent rule   **D.** polarity rule   **E.** none of the above

95. How many ions are produced in solution by dissociation of one formula unit of $Co(NO_3)_2 \cdot 6H_2O$?

    A. 2   **B.** 3   **C.** 4   **D.** 6   **E.** 9

96. When a solid dissolves, each molecule is removed from the crystal by interaction with the solvent. This process of surrounding each ion with solvent molecules is called:

    A. hemolysis   **C.** crenation

    B. electrolysis   **D.** dilution   **E.** solvation

97. Which of the following represents the bicarbonate ion?

    A. HCO_3^- **B.** $H_2CO_3^{2-}$ **C.** CO_3^- **D.** CO_3^{2-} **E.** HCO_3^{2-}

98. Coca-Cola is carbonated by injection with carbon dioxide gas. Under what conditions is carbon dioxide gas most soluble?

    A. high temperature, high pressure   **C.** low temperature, low pressure

    B. high temperature, low pressure   **D.** low temperature, high pressure

                                               E. none of the above

99. What is the concentration of ^-OH if the concentration of H_3O^+ is 1×10^{-5} M?

    $[H_3O^+] \times [^-OH] = K_w = 1 \times 10^{-14}$

    A. 1×10^5 **B.** 1×10^9 **C.** 1×10^{-5} **D.** 1×10^{-14} **E.** 1×10^{-9}

100. What is the concentration in mass-volume percent for 13.9 g CaF_2 in 255 mL of solution?

    A. 5.45%   **B.** 0.950%   **C.** 0.180%   **D.** 2.73%   **E.** 0.360%

101. Soft drinks are carbonated by injection with carbon dioxide gas. Under what conditions is carbon dioxide gas least soluble?

    A. high temperature, low pressure   **C.** low temperature, high pressure

    B. high temperature, high pressure   **D.** low temperature, low pressure

                                               E. none of the above

102. Which of the following can be classified as a colloid?

    A. Grape juice   **C.** Pepsi-Cola

    B. Homogenized milk   **D.** Kool-Aid   **E.** Decaf coffee

103. Which ions are the spectator ions in the reaction below?

    $K_2SO_4\,(aq) + Ba(NO_3)_2\,(aq) \rightarrow BaSO_4\,(s) + 2KNO_3\,(aq)$?

    A. K^+ and SO_4^{2-} **C.** Ba^{2+} and K^+

    B. Ba^{2+} and NO_3^- **D.** Ba^{2+} and SO_4^{2-} **E.** K^+ and NO_3^-

                                            Copyright © 2015 Sterling Test Prep

104. A substance represented by a formula written as $M_xLO_y \cdot zH_2O$ is called a:

A. solvent
C. solid hydrate
B. solute
D. colloid
E. suspension

105. Which is the correct order for nitrate, nitrite, sulfate and sulfite?

A. NO_2^-, NO_3^-, SO_4^-, SO_3^-
C. NO_3^-, NO_2^-, SO_3^-, SO_4^-
B. NO_3^-, NO_2^-, SO_4^-, SO_3^-
D. NO_2^-, NO_3^-, SO_3^-, SO_4^-
E. NO_2^-, SO_3^-, NO_3^-, SO_4^-

106. If the solubility of nitrogen in blood is 1.90 cc/100 cc at 1.00 atm, what is the solubility of nitrogen in a scuba diver's blood at a depth of 150 feet where the pressure is 5.55 atm?

A. 1.90 cc/100 cc
C. 0.190 cc/100 cc
B. 3.80 cc/100 cc
D. 0.380 cc/100 cc
E. 10.5 cc/100 cc

107. What is the concentration of ^-OH, if the concentration of H_3O^+ is 1×10^{-8} M?

$$[H_3O^+] \times [^-OH] = K_w = 1 \times 10^{-14}$$

A. 1×10^8 **B.** 1×10^6 **C.** 1×10^{-14} **D.** 1×10^{-6} **E.** 1×10^{-8}

108. What volume of 8.50% (m/v) solution contains 50.0 grams of glucose?

A. 17.0 mL   **B.** 48.0 mL   **C.** 588 mL   **D.** 344 mL   **E.** 96.0 mL

109. What is the solubility of CaF_2 when it is added to a 0.02 M solution of NaF?

A. Low because $[F^-]$ present in solution inhibits dissociation of CaF_2
B. High because of the common ion effect
C. Unaffected by the presence of NaF
D. Low because less water is available for solvation due to the presence of NaF
E. High because $[F^-]$ present in solution facilitates dissociation of CaF_2

110. Which of the following illustrates the *like dissolves like* rule for two liquids?

A. A polar solute is miscible with a polar solvent
B. A polar solute is immiscible with a nonpolar solvent
C. A nonpolar solute is miscible with a nonpolar solvent
D. All of the above
E. None of the above

111. If X moles of $PbCl_2$ fully dissociate in 1 liter of H_2O, the K_{sp} product is equivalent to:

A. X^2 **B.** $2X^4$ **C.** $3X^2$ **D.** $2X^3$ **E.** $4X^3$

112. Which of the following concentrations is dependent on temperature?

A. mole fraction
C. mass percent
B. molarity
D. molality
E. more than one of the above

113. An ionic compound that strongly attracts atmospheric water is said to be:

A. immiscible   **B.** miscible   **C.** diluted   **D.** hygroscopic   **E.** soluble

114. Hydration involves the:

A. formation of bonds between the solute and water molecules

B. breaking of water–water bonds

C. breaking of water–solute bonds

D. breaking of solute–solute bonds

E. both the breaking of water-water bonds and formation of solute-solute bonds

115. Which of the following illustrates the *like dissolves like* rule for two liquids?

A. A nonpolar solute is miscible with a nonpolar solvent

B. A nonpolar solvent is miscible with a polar solvent

C. A polar solute is miscible with a nonpolar solvent

D. A polar solute is immiscible with a polar solvent

E. A polar solvent is immiscible with a polar solute

116. With increasing temperature, many solvents expand to occupy greater volumes. What happens to the concentration of a solution made with such a solvent as temperature increases?

A. Concentration decreases because the solution has a greater ability to dissolve more solute at a higher temperature

B. Concentration increases because the solution has a greater ability to dissolve more solute at a higher temperature

C. Concentration of a solution decreases as the volume increases because concentration depends on how much mass is dissolved in a given volume

D. Concentration of a solution increases as the solute fits into the new spaces between the molecules

E. Concentration of a solution increases as the volume increases because concentration depends on how much mass is dissolved in a given volume

117. What mass of H_2O is needed to prepare 148 grams of 12.0% (m/m) $KHCO_3$ solution?

A. 120 g   **B.** 130 g   **C.** 11.3 g   **D.** 17.8 g   **E.** 12.0 g

118. Which of the following is an example of a true solution?

A. Apple juice   **C.** Homogenized milk

B. Mayonnaise   **D.** Blood   **E.** Glass of water with ice

119. The *like dissolves like* rule for two liquids is illustrated by which of the following?

A. A nonpolar solute is immiscible with a nonpolar solvent

B. A nonpolar solute is miscible with a polar solvent

C. A polar solute is immiscible with a polar solvent

D. A polar solute is miscible with a nonpolar solvent

E. None of the above

Copyright © 2015 Sterling Test Prep

120. What is the concentration of Ag in moles/liter if the K_{sp} for AgCl is A, and the concentration Cl⁻ in a container is B molar?

    A. A moles/liter   **C.** A/B moles/liter

    B. B moles/liter   **D.** B or C   **E.** none of the above

121. In the reaction KHS (*aq*) + HCl (*aq*) → KCl (*aq*) + H_2S (*g*), which ions are the spectator ions?

    A. H^+ and HS⁻   **B.** K^+ and HS⁻   **C.** K^+ and Cl⁻   **D.** K^+ and H^+ **E.** HS⁻ and Cl⁻

122. A supersaturated solution:

    A. contains a dissolved solute in equilibrium with an undissolved solid

    B. will rapidly precipitate if a seed crystal is added

    C. contains as much solvent as it can hold

    D. contains no double bonds

    E. none of the above

123. Apply the *like dissolves like* rule to predict which of the following liquids is miscible with water.

    A. methyl ethyl ketone, C_4H_8O **C.** formic acid, $HCHO_2$

    B. glycerin, $C_3H_5(OH)_3$ **D.** all of the above   **E.** none of the above

124. The hydration number of an ion is the number of:

    A. water molecules bonded to an ion in an aqueous solution

    B. water molecules required to dissolve one mole of ions

    C. ions bonded to one mole of water molecules

    D. ions dissolved in one liter of an aqueous solution

    E. water molecules required to dissolve the compound

125. How many grams of sucrose are in 5.0 liters of sugar water that has a concentration of 0.50 grams / liter of solution?

    A. 0.5 g   **B.** 5.0 g   **C.** 2.5 g   **D.** 1.0 g   **E.** 1.5 g

126. What mass of NaOH is contained in 75.0 mL of a 5.0% (m/v) NaOH solution?

    A. 6.50g   **B.** 15.0g   **C.** 7.50g   **D.** 0.65g   **E.** 3.75g

127. Apply the like dissolves like rule to predict which of the following liquids is miscible with water.

    A. carbon tetrachloride, CCl_4 **C.** ethanol, C_2H_5OH

    B. toluene, C_7H_8 **D.** all of the above   **E.** none of the above

128. A water solution of sodium chloride is a good conductor of electricity. Therefore, table salt is classified as a(n):

    A. non–electrolyte   **C.** weak electrolyte

    B. semi–electrolyte   **D.** strong electrolyte   **E.** strong ionic compound

129. Which substance does not produce ions when dissolved in water?

   A. manganese (II) nitrate   **C.** CH_2O

   B. CsCN   **D.** KClO   **E.** NaHS

130. Apply the *like dissolves like* rule to predict which of the liquids is immiscible in water.

   A. ethanol, CH3CH₂OH   **C.** acetic acid, CH_3COOH

   B. acetone, C_3H_6O **D.** all of the above   **E.** none of the above

131. Which of the following is/are a strong electrolytes?

       I. salts   II. strong bases   III. weak acids

   A. I only   **C.** III only

   B. I and II only   **D.** I, II and III   **E.** I and III only

132. How are intermolecular forces and solubility related?

   A. Solubility is a measure of how weak the intermolecular forces in the solute are

   B. Solubility is a measure of how strong a solvent's intermolecular forces are

   C. Solubility depends on the solute's ability to overcome the intermolecular forces in the solvent

   D. Solubility depends on the solvent's ability to overcome the intermolecular forces in a solid

   E. None of the above

133. Calculate the mass percent of a solution prepared by dissolving 17.2 g of NaCl in 149 g of water:

   A. 10.4%   **B.** 0.103%   **C.** 12.4%   **D.** 6.45%   **E.** 16.9%

134. Apply the like dissolves like rule to predict which of the following liquids is miscible with liquid bromine, Br_2.

   A. benzene (C_6H_6)   **C.** carbon tetrachloride (CCl_4)

   B. hexane (C_6H_{14})   **D.** none of the above   **E.** all of the above

135. What is the concentration of I⁻ ions in a 0.40 M solution of magnesium iodide?

   A. 0.05 M   **B.** 0.80 M   **C.** 0.60 M   **D.** 0.20 M   **E.** 0.40 M

136. Calculate the [N_2] gas in water, if N_2 gas at a partial pressure above the water is 0.826 atm and Henry's Law constant for N_2 in water = 6.8×10^{-4} mol/L-atm.

   A. 5.6×10^{-4} M   **C.** 3.2×10^3 M

   B. 7.2×10^{-3} M   **D.** 0.61 M   **E.** 7.2×10^{-5} M

137. What was the molarity of the original solution if a 20.0 mL sample of a $CuSO_4$ solution was dried and 0.967 g of copper (II) sulfate remained?

   A. 0.606 M   **B.** 0.570 M   **C.** 0.253 M   **D.** 0.303 M   **E.** 0.909 M

                                             Copyright © 2015 Sterling Test Prep

Use the following data to answer questions **138** through **142**.

$$K_{sp} \quad PbCl_2 = 1.0 \times 10^{-5}$$
$$K_{sp} \quad AgCl = 1.0 \times 10^{-10}$$
$$K_{sp} \quad PbCO_3 = 1.0 \times 10^{-15}$$

138. Comparing equal volumes of saturated solutions for $PbCl_2$ and AgCl, which solution contains a greater concentration of Cl^-?

      I. $PbCl_2$

      II. AgCl

      III. Both have the same concentration of Cl^-

  A. I only   **B.** II only   **C.** III only   **D.** I, II and III   **E.** I and II only

139. Considering saturated solutions for $PbCl_2$ and AgCl, is Ag^+ or Pb^{2+} in a higher concentration?

      I. Ag^+

      II. AgCl

      III. Both have the same concentration

  A. I only   **B.** II only   **C.** III only   **D.** I, II and III   **E.** I and II only

140. Consider a saturated solution of $PbCl_2$. The addition of NaCl would:

  A. decrease $[Pb^{2+}]$ **C.** increase the precipitation of $PbCl_2$

  B. have no affect on the precipitation of $PbCl_2$ **D.** A and B

                                               E. A and C

141. What occurs when $AgNO_3$ is added to a saturated solution of $PbCl_2$?

      I. AgCl precipitates

      II. $Pb(NO_3)_2$ forms a white precipitate

      III. More $PbCl_2$ forms

  A. I only   **B.** II only   **C.** III only   **D.** I, II and III   **E.** I and II only

142. If the K_{sp} of $MgSO_4$ is 4×10^{-5}, would a precipitate form if 1 liter of 0.03M $Mg(NO_3)_2$ were mixed with 2 liters of 0.06 M K_2SO_4? (Assume complete dissociation of solutions.)

  A. Yes

  B. No, because the K_{sp} value for $MgSO_4$ is not exceeded

  C. No, because the solution does not contain $MgSO_4$

  D. Not enough data is given

  E. No, because the K_{sp} value for $MgSO_4$ is exceeded

143. Which of the following illustrates the *like dissolves like* rule for a solid solute in a liquid solvent?

      I. A nonpolar compound is soluble in a nonpolar solvent

      II. A polar compound is soluble in a polar solvent

      III. An ionic compound is soluble in a polar solvent

  A. I only   **B.** II only   **C.** I and II only   **D.** I, II and III   **E.** None of the above

144. Potassium oxide contains:

A. ionic bonds and is a weak electrolyte

B. ionic bonds and is a strong electrolyte

C. covalent bonds and is a non-electrolyte

D. covalent bonds and is a strong electrolyte

E. hydrogen bonds and is a non-electrolyte

145. It is determined that 9.86×10^{-3} g of a contaminant is present in 4865 g of a particular solution. What is the concentration of the contaminant in ppm (m/m)?

A. 3.40 ppm   B. 1.50 ppm   C. 4.06 ppm   D. 2.03 ppm   E. 1.70 ppm

146. Which of the following illustrates the *like dissolves like* rule for a solid solute in a liquid solvent?

A. An ionic compound is soluble in a nonpolar solvent

B. A nonpolar compound is soluble in a polar solvent

C. A polar compound is soluble in a nonpolar solvent

D. All of the above

E. None of the above

147. What happens when the molecule-to-molecule attractions in the solute are comparable to those in the solvent?

A. The material has only limited solubility in the solvent

B. The solution will become saturated

C. The solute can have infinite solubility in the solvent

D. The solute does not dissolve in the solvent

E. None of the above

148. A water solution of the table sugar sucrose does not conduct electricity. Due to this property, sucrose is classified as:

A. semi-electrolyte

B. non-electrolyte

C. strong electrolyte

D. weak electrolyte

E. neither a strong nor weak electrolyte

149. Calculate the molarity of a 9.55 molal solution of methanol (CH_3OH) with a density of 0.937 g/Ml:

A. 0.155 M   B. 23.5 M   C. 6.86 M   D. 68.6 M   E. 2.35 M

150. From the *like dissolves like* rule, predict which of the following vitamins is soluble in water.

A. α-tocopherol ($C_{29}H_{50}O_2$)

B. calciferol ($C_{27}H_{44}O$)

C. ascorbic acid ($C_6H_8O_6$)

D. retinol ($C_{20}H_{30}O$)

E. none of the above

151. The equation for the reaction $BaCl_2\,(aq) + K_2CrO_4\,(aq) \rightarrow BaCrO_4\,(s) + KCl\,(aq)$ can be written as an ionic equation. In the ionic equation, the spectator ions are:

A. K^+ and Cl^-

B. B^{2+} and CrO_4^{2-}

C. Ba^{2+} and K^+

D. K^+ and CrO_4^{2-}

E. Cl^- and CrO_4^{2-}

Copyright © 2015 Sterling Test Prep

152. Which solution is the least concentrated? Each choice refers to the same solute and solvent.

- **A.** 2.4 g solute in 2 mL solution
- **B.** 2.4 g solute in 5 mL solution
- **C.** 20 g solute in 50 mL solution
- **D.** 30 g solute in 150 mL solution
- **E.** 50 g solute in 175 mL solution

153. Which of the following is true of a colloid?

- **A.** Dispersed particles are less than 1 nm in diameter
- **B.** Dispersed particles demonstrate the Tyndall effect
- **C.** Dispersed particles pass through biological membranes
- **D.** All of the above
- **E.** None of the above

154. Hydrochloric acid contains:

- **A.** ionic bonds and is a strong electrolyte
- **B.** hydrogen bonds and is a weak electrolyte
- **C.** covalent bonds and is a strong electrolyte
- **D.** ionic bonds and is a non-electrolyte
- **E.** covalent bonds and is a non-electrolyte

155. Apply the *like dissolves like* rule to predict which of the following vitamins is insoluble in water.

- **A.** niacinamide ($C_6H_6N_2O$)
- **B.** pyridoxine ($C_8H_{11}NO_3$)
- **C.** retinol ($C_{20}H_{30}$)
- **D.** thiamine ($C_{12}H_{17}N_4OS$)
- **E.** cyanocobalamin ($C_{63}H_{88}CoN_{14}O_{14}P$)

156. Molarity could be used as a conversion factor between:

- **A.** grams of solute and volume of solution
- **B.** moles of solute and kilograms of solution
- **C.** grams of solute and moles of solvent
- **D.** moles of solute and volume of solution
- **E.** grams of solute and kilograms of solution

157. Which of the following might have the best solubility in water?

- **A.** CH_3CH_3 **B.** CH_3OH **C.** Cl_2 **D.** O_2 **E.** none of the above

158. If 10.0 mL of blood plasma has a mass of 10.279 g and contains 0.870 g of protein, what is the mass/mass percent concentration of protein in the blood plasma?

- **A.** 8.70%   **B.** 32.1%   **C.** 0.870%   **D.** 97.3%   **E.** 8.46%

159. At 20 °C, a one-liter sample of pure water has a vapor pressure of 17.5 torr. If 5 g of NaCl is added to the sample, the vapor pressure of the water:

- **A.** equals the vapor pressure of the added NaCl
- **B.** remain unchanged at 17.5 torr
- **C.** increases
- **D.** decreases
- **E.** increases to the square root of solute added

160. Which of the following is an example of a weak electrolyte?

- **A.** H_2SO_4 **B.** HNO_3 **C.** $HC_2H_3O_2$ **D.** $MgCl_2$ **E.** Na_2CO_3

161. If 25.0 mL of urine has a mass of 25.725 g and contains 1.929 g of solute, what is the mass/mass percent concentration of solute in the urine sample?

A. 13.34%   **B.** 7.72%   **C.** 7.499%   **D.** 3.887%   **E.** 97.2%

162. Which information is necessary to determine the molarity of a solution if the chemical formula of the solute is known?

   A. Mass of the solute dissolved and the volume of the solvent added
   B. Molar mass of both the solute and the solvent used
   C. Only the volume of the solvent used
   D. Only the mass of the solute dissolved
   E. Mass of the solute dissolved and the final volume of the solution

163. In the reaction between aqueous silver nitrate and aqueous potassium chromate, what is the identity of the soluble substance that is formed?

   A. potassium nitrate
   B. potassium chromate
   C. silver nitrate
   D. silver chromate
   E. no soluble substance is formed

164. If 25.0 mL of seawater has a mass of 25.895 g and contains 1.295 g of solute, what is the mass/mass percent concentration of solute in the seawater sample?

A. 5.18%   **B.** 20.00%   **C.** 3.862%   **D.** 5.001%   **E.** 96.5%

165. When salt A is dissolved into water to form a 1 molar unsaturated solution, the temperature of the solution decreases. Under these conditions, which statement is accurate when salt A is dissolved in water?

   A. $\Delta H°$ and $\Delta G°$ are positive
   B. $\Delta H°$ is positive and $\Delta G°$ is negative
   C. $\Delta H°$ is negative and $\Delta G°$ is positive
   D. $\Delta H°$ and $\Delta G°$ are negative
   E. $\Delta H°$, $\Delta S°$ and $\Delta G°$ are positive

166. Under which conditions is the expected solubility of oxygen gas in water the highest?

   A. High temperature and high O_2 pressure above the solution
   B. Low temperature and low O_2 pressure above the solution
   C. Low temperature and high O_2 pressure above the solution
   D. High temperature and low O_2 pressure above the solution
   E. The O_2 solubility is independent of temperature and pressure

167. How many grams of H_3PO_4 are needed to make 175 mL of a 0.175 M H_3PO_4 solution?

A. 3.00 g   **B.** 5.08 g   **C.** 0.346 g   **D.** 0.764 g   **E.** 2.54 g

168. Which of the following is NOT a unit factor related to a 15.0% aqueous solution of potassium iodide (KI)?

   A. 100 g solution / 85.0 g water
   B. 85.0 g water / 15.0 g KI
   C. 85.0 g water / 100 g solution
   D. 15.0 g KI/ 85.0 g water
   E. 15.0 g KI / 100 g water

Copyright © 2015 Sterling Test Prep

169. At room temperature, barium hydroxide ($Ba(OH)_2$) is dissolved in pure water. From the equilibrium concentrations of Ba^+ and OH^- ions, which expressions could be used to find the solubility product for barium hydroxide?

    A. $K_{sp} = [Ba][OH]^2$ **C.** $K_{sp} = [Ba][OH]$

    B. $K_{sp} = 2[Ba][OH]^2$ **D.** $K_{sp} = [Ba]2[OH]$ **E.** $K_{sp} = 2[Ba][OH]$

170. Which of the following is an example of a nonelectrolyte?

    A. Na_2SO_4 **B.** NaCl   **C.** $MgCl_2$ **D.** $HC_2H_3O_2$ **E.** $C_{12}H_{22}O_{11}$

171. What is the mass of a 10.0% blood plasma sample that contains 2.50 g of dissolved solute?

    A. 21.5 g   **B.** 25.0 g   **C.** 0.215 g   **D.** 0.430 g   **E.** 12.5 g

172. Which of the following would likely form micelles in an aqueous solution?

    A. dodecanoic acid   **C.** glucose

    B. glutamic acid   **D.** hexane   **E.** none of the above

173. The heat of a solution measures the energy absorbed from:

    I. formation of solvent-solute bonds

    II. breaking of solute-solute bonds

    III. breaking of solvent-solvent bonds

    A. I only   **C.** III only

    B. II only   **D.** I, II and III   **E.** I and II only

174. What is the mass of a 7.50% urine sample that contains 122 g of dissolved solute?

    A. 1550 g   **C.** 9.35 g

    B. 935 g   **D.** 15.5 g   **E.** 1630 g

175. Which statement below is true?

    A. All bases are strong electrolytes and ionize completely when dissolved in water

    B. All salts are strong electrolytes and dissociate completely when dissolved in water

    C. All acids are strong electrolytes and ionize completely when dissolved in water

    D. All bases are weak electrolytes and ionize completely when dissolved in water

    E. All salts are weak electrolytes and ionize partially when dissolved in water

176. What is the molarity of KCl in sea water if sea water is 12.5% (m/m) and the density of sea water is 1.06 g/mL?

    A. 1.78 M   **B.** 17.8 M   **C.** 2.78 M   **D.** 0.845 M   **E.** 27.8 M

177. What is the molarity of a glucose solution that contains 10.0 g of $C_6H_{12}O_6$ (180.18 g/mol) dissolved in 100.0 mL of solution?

    A. 1.80 M   **C.** 0.0555 M

    B. 0.555 M   **D.** 0.00555 M   **E.** 18.0 M

178. Why does the reaction proceed if, when solid potassium chloride is dissolved in H_2O, the energy of the bonds formed is less than the energy of the bonds broken?

    A. The electronegativity of the H_2O increases from interaction with potassium and chloride ions

    B. The reaction does not take place under standard conditions

    C. The decreased disorder due to mixing decreases entropy within the system

    D. Remaining potassium chloride which does not dissolve offsets the portion that dissolves

    E. The increased disorder due to mixing increases entropy within the system

179. Which statement supports the fact that calcium fluoride is much less soluble in water than sodium fluoride?

    I. calcium fluoride is not used in toothpaste

    II. calcium fluoride is not used to fluoridate city water supplies

    III. sodium fluoride is not used to fluoridate city water supplies

    A. I only   **C.** III only

    B. II only   **D.** I and II only   **E.** I, II and III

180. What is the molarity of a sucrose solution that contains 10.0 g of $C_{12}H_{22}O_{11}$ (342.34 g/mol) dissolved in 100.0 mL of solution?

    A. 0.292 M   **C.** 0.00292 M

    B. 3.33 M   **D.** 0.0292 M   **E.** 33.3 M

181. What is the molar concentration of Ca^{2+} (*aq*) in a solution that is prepared by mixing 15 mL of a 0.02 M $CaCl_2$ (*aq*) solution with a 10 mL of a 0.04 M $CaSO_4$ (*aq*) solution?

    A. 0.014 M   **C.** 0.028 M

    B. 0.020 M   **D.** 0.042 M   **E.** 0.046 M

182. Which is a correctly balanced hydration equation for the hydration of Na_2SO_4?

    A. $Na_2SO_4\ (s) \xrightarrow{H_2O} Na^+\ (aq) + 2SO_4^{2-}\ (aq)$

    B. $Na_2SO_4\ (s) \xrightarrow{H_2O} 2Na^{2+}\ (aq) + S^{2-}\ (aq) + O_4^{2-}\ (aq)$

    C. $Na_2SO_4\ (s) \xrightarrow{H_2O} Na_2^{2+}\ (aq) + SO_4^{2-}\ (aq)$

    D. $Na_2SO_4\ (s) \xrightarrow{H_2O} 2Na^+\ (aq) + SO_4^{2-}\ (aq)$

    E. $Na_2SO_4\ (s) \xrightarrow{H_2O} 2Na^{2+}\ (aq) + S^{2-}\ (aq) + SO_4^{2-}\ (aq) + O_4^{2-}\ (aq)$

183. What is the molarity of a hydrochloric acid solution prepared by diluting 250.0 mL of 6.00 M HCl to a total volume of 2.50 L?

    A. 2.50 M   **C.** 0.250 M

    B. 0.600 M   **D.** 0.0600 M   **E.** 6.00 M

Copyright © 2015 Sterling Test Prep

184. What happens to DNA when placed into an aqueous solution at physiological pH?

  A. Individual DNA molecules repel each other due to presence of positive charges

  B. DNA molecules bind to negatively charged proteins

  C. Individual DNA molecules attract each other due to presence of positive and negative charges

  D. Individual DNA molecules repel each other due to presence of negative charges

  E. None of the above

185. What is the molarity of a solution prepared by dissolving 3.50 mol NaCl in enough water to make 1.50 L of solution?

  A. 0.429 M   **B.** 2.33 M   **C.** 5.25 M   **D.** 87.8 M   **E.** 137 M

186. What is the molarity of a hydrochloric acid solution prepared by diluting 200.0 mL of 0.500 M HCl to a total volume of 1.00 L?

  A. 0.250 M   **B.** 1.00 M   **C.** 0.100 M   **D.** 0.200 M   **E.** 2.50 M

187. What does a negative heat of solution indicate about solute-solvent bonds as compared to solute-solute bonds and solvent-solvent bonds?

  A. Solute-solute and solute-solvent bond strengths are greater than solvent-solvent bond strength

  B. Heat of solution does not support conclusion about bond strength

  C. Solute-solute and solvent-solvent bonds are stronger than solute-solvent bonds

  D. Solute-solute and solvent-solvent bond strengths are equal to solute-solvent bond strength

  E. Solute-solute and solvent-solvent bonds are weaker than solute-solvent bonds

188. What is the formula of the solid formed when aqueous barium chloride is mixed with aqueous potassium chromate?

  A. K_2CrO_4 **B.** K_2Ba **C.** KCl   **D.** $BaCrO_4$ **E.** $BaCl_2$

189. What is the molarity of a hydrochloric acid solution prepared by diluting 500.0 mL of 1.00 M HCl to a total volume of 2.50 L?

  A. 2.00 M   **B.** 1.00 M   **C.** 2.50 M   **D.** 0.100 M   **E.** 0.200 M

190. Why might sodium carbonate (washing soda, Na_2CO_3) be added to hard water for cleaning?

  A. The soap gets softer due to the added ions

  B. The ions solubilize the soap due to ion-ion intermolecular attraction, which improves the cleaning ability

  C. The hard ions in the water are more attracted to the carbonate ions -2 charge

  D. The hard ions are dissolved by the added sodium ions

  E. None of the above

191. What mass of KOH is needed to produce 22.0 mL of 0.576 M solution?

  A. 2.20 g   **B.** 0.711 g   **C.** 1.48 g   **D.** 0.443 g   **E.** 4.43 g

192. What volume of 12 M acid must be diluted with distilled water to prepare 5.0 L of 0.10 M acid?

A. 42 mL　　**B.** 60 mL　　**C.** 0.042 mL　　**D.** 6 mL　　**E.** 420 mL

193. What is the %v/v concentration of red wine that contains 15 mL of ethyl alcohol in 200 mL?

A. 0.60 %v/v　　　　**C.** 0.075 %v/v

B. 0.060 %v/v　　　　**D.** 6.0 %v/v　　　　**E.** 7.5 %v/v

194. If 29.4 g of LiOH is dissolved in water to make 985 mL of solution, what is the molarity of the LiOH solution?

A. 1.25 M　　**B.** 2.29 M　　**C.** 0.986 M　　**D.** 0.478 M　　**E.** 0.229 M

195. What volume of 0.255 M hydrochloric acid reacts completely with 0.400 g of sodium hydrogen carbonate, $NaHCO_3$ (84.01 g/mol) ?

$$NaHCO_3 \, (s) + HCl \, (aq) \rightarrow NaCl \, (aq) + H_2O \, (l) + CO_2 \, (g)$$

A. 121 mL　　**B.** 53.6 mL　　**C.** 18.7 mL　　**D.** 4.76 mL　　**E.** 132 mL

196. All of the statements about molarity are correct EXCEPT:

A. volume = moles/molarity

B. moles = molarity × volume

C. molarity of a diluted solution is less than the molarity of the original solution

D. abbreviation is M

E. interpretation of the symbol is "moles of solute per mole of solvent"

197. What is the v/v% concentration of a solution made by adding 25 mL of acetone to 75 mL of water?

A. 33 % v/v　　**B.** 0.33 % v/v　　**C.** 25 % v/v　　**D.** 0.25 % v/v　　**E.** 3.3 % v/v

198. What volume of 0.115 M hydrochloric acid reacts completely with 0.125 g of sodium carbonate, Na_2CO_3 (105.99 g/mol)?

$$Na_2CO_3 \, (s) + 2HCl \, (a) \rightarrow Na_2SO_4 \, (aq) + 2H_2O \, (l) + 2CO_2 \, (g)$$

A. 20.5 mL　　**B.** 41.0 mL　　**C.** 9.75 mL　　**D.** 10.3 mL　　**E.** 97.5 mL

199. What mass of $CaCl_2$ is required to prepare 3.20 L of 0.850 M $CaCl_2$ solution?

A. 75.5 g　　**B.** 160 g　　**C.** 320 g　　**D.** 302 g　　**E.** 151 g

200. What is the net ionic equation for the reaction between HNO_3 and Na_2SO_3?

A. $H_2O + SO_2 \, (g) \rightarrow H^+ \, (aq) + HSO_3^- \, (aq)$

B. $2H^+ \, (aq) + SO_3^{2-} \, (aq) \rightarrow H_2O + SO_2 \, (g)$

C. $HNO_3 \, (aq) + Na_2SO_3 \, (aq) \rightarrow H_2O + SO_2 \, (g) + NO_3^- \, (aq)$

D. $H_2SO_3 \, (aq) \rightarrow H_2O + SO_2 \, (g)$

E. $2H^+ \, (aq) + Na_2SO_3 \, (aq) \rightarrow 2Na^+ \, (aq) + H_2O + SO_2 \, (g)$

201. What solution conditions are required for a protein to be a positively charged macroion?

  A. pH of solution is less than the protein's pI and ionic strength is low
  B. pH of solution is greater than the protein's pI and ionic strength is low
  C. pH of solution is less than the protein's pI
  D. pH of solution is greater than the protein's pI
  E. None of the above

202. Which of the following symbolizes a liquid in a chemical equation?

  A. (*aq*)　　　**B.** (*g*)　　　**C.** (*l*)　　　**D.** (*s*)　　　**E.** none of the above

203. How many grams of $FeSO_4$ are present in a 20.0 mL sample of a 0.500 M solution?

  A. 1.52 g　　　**B.** 0.760 g　　　**C.** 60.4 g　　　**D.** 6.04 g　　　**E.** 7.60 g

204. Which is the net ionic equation for the reaction that takes place when $HC_2H_3O_2$ (*aq*) is added to NH_3 (*aq*)?

  A. $HC_2H_3O_2$ (*aq*) + OH^- (*aq*) → $C_2H_3O_2^-$ (*aq*) + H_2O (*l*)
  B. H^+ (*aq*) + NH_4OH (*aq*) → NH_4^+ (*aq*) + H_2O (*l*)
  C. $HC_2H_3O_2$ (*aq*) + NH_4OH (*aq*) → H_2O (*l*) + NH_4ClO_4 (*aq*)
  D. H^+ (*aq*) + OH^- (*aq*) → H_2O (*l*)
  E. $HC_2H_3O_2$ (*aq*) + NH_3 (*aq*) → NH_4^+ (*aq*) + $C_2H_3O_2^-$ (*aq*)

205. Which of the following statements best describes what is happening in a water softening unit?

  A. The sodium is removed from the water, making the water interact less with the soap molecules
  B. The ions in the water softener are softened by chemically bonding with sodium
  C. The hard ions are all trapped in the softener, which filters out all the ions
  D. The hard ions in water are exchanged for ions that do not interact as strongly with soaps
  E. None of the above

206. Which of the following explains why bubbles form on the inside of a pan of water when the pan of water is heated?

  A. As temperature increases, the vapor pressure increases
  B. As temperature increases, the atmospheric pressure decreases
  C. As temperature increases, the solubility of air decreases
  D. As temperature increases, the kinetic energy decreases
  E. None of the above

207. The density of a $NaNO_3$ solution is 1.24 g/mL. What is the molarity of a 25.0% (m/m) $NaNO_3$ solution?

  A. 15.3 M　　**B.** 8.41 M　　**C.** 0.760 M　　**D.** 3.65 M　　**E.** 7.60 M

208. Which is the net ionic equation for the reaction when HNO_3 (aq) is added to Fe_2O_3 (s)?

A. $6H^+$ $(aq) + Fe_2O_3$ $(s) \rightarrow 2Fe^{3+}$ $(aq) + 3H_2O$ (l)

B. HNO_3 $(aq) + OH^-$ $(s) \rightarrow NO_3^-$ $(aq) + H_2O$ (l)

C. H^+ $(aq) + OH^-$ $(aq) \rightarrow H_2O$ (l)

D. HNO_3 $(aq) + Fe_2O_3$ $(s) \rightarrow Fe(NO_3)_3$ $(aq) + H_2O$ (l)

E. HNO_3 $(aq) + Fe^{3+}$ $(aq) \rightarrow Fe(NO_3)_3$ $(aq) + H^+$ (aq)

209. How many grams of NaOH are needed to make 750 mL of a 2.5% (w/v) solution?

A. 2.4 g   **B.** 7.5 g   **C.** 19 g   **D.** 15 g   **E.** 1.5 g

210. If the solubility of carbon dioxide in a bottle of champagne is 1.45 g per liter at 1.00 atmosphere, what is the solubility of carbon dioxide at 10.0 atmospheres?

A. 3.40 g/L   **B.** 1.25 g/L   **C.** 0.720 g/L   **D.** 0.145 g/L   **E.** 14.5 g/L

211. Which of the following indicates an aqueous solution of a substance in a chemical equation?

A. (g) **B.** (aq) **C.** (s) **D.** (l) **E.** none of the above

Copyright © 2015 Sterling Test Prep

Chapter 5.   ACIDS & BASES

1. Which of the following is a strong acid?

   A. H_3PO_4 (*aq*) (~1% ionized)

   B. H_2SO_3 (*aq*) (~1% ionized)

   C. HNO_2 (*aq*) (~1% ionized)

   D. all of the above

   E. none of the above

2. Which of the following properties is NOT characteristic of an acid?

   A. Is neutralized by a base

   B. Has a slippery feel

   C. Produces H^+ in water

   D. Has a sour taste

   E. Has a pH reading less than 7

3. When NH_3 is diluted in H_2O, the reaction below occurs. What is the concentration of the ammonium ion, if the concentration of ammonia in a solution is 0.4 M and the pH = 10?

$$NH_3 + H_2O \rightleftharpoons NH_4^+ + OH^- \qquad K_b = 2 \times 10^{-5}$$

   A. 0.01 M   **B.** 0.02 M   **C.** 0.08 M   **D.** 0.04 M   **E.** 0.10 M

4. Which of the following is an example of an Arrhenius base?

   A. NaOH (*aq*)

   B. Al $(OH)_3$ (*s*)

   C. $Ca(OH)_2$ (*aq*)

   D. all of the above

   E. None of the above

5. Which of the following substances is produced during an acid/base (i.e. neutralization) reaction?

   A. NaOH   **B.** CO_2 **C.** H_2O **D.** H_2 **E.** NaCl

6. What atom in the ammonium ion (NH_4^+) bears the positive charge?

   A. Nitrogen atom

   B. Hydrogen atom

   C. Neither atom bears a positive charge

   D. Nitrogen and hydrogen equally share the positive charge

   E. The spectator ion holds the positive charge

7. When dissolved in water, the Arrhenius acid-bases KOH, H_2SO_4 and HNO_3 are, respectively:

   A. base, acid and base

   B. base, base and acid

   C. base, acid and acid

   D. acid, base and base

   E. acid, acid and base

8. What are the predicted products from the following neutralization reaction?

$$HCl\ (aq) + NH_4OH\ (aq) \rightarrow$$

   A. NH_4Cl and H_2O

   B. NH_4Cl and O_2

   C. NH_3Cl and H_2O

   D. NH_3Cl and O_2

   E. No reaction

9. Which of the following is a weak acid?

   A. OH^- **B.** NH_3 **C.** HNO_3 **D.** HF   **E.** HI

10. A Brønsted-Lowry base is a(n):

   A. electron acceptor   **C.** electron donor

   B. proton acceptor   **D.** proton donor

                             E. both proton donor and electron acceptor

11. Which of the following pairs of 0.1 M solutions react to form a precipitate?

   A. KOH and $Ba(NO_3)_2$ **C.** K_2SO_4 and CsI

   B. NaI and KBr   **D.** $NiBr_2$ and $AgNO_3$

                             E. all of the above

12. What are products from the neutralization reaction $HC_2H_3O_2\,(aq) + Ca(OH)_2\,(aq) \rightarrow$?

   A. $Ca(C_2H_3O_2)_2$ and H_2 **C.** $Ca(HCO_3)_2$ and H_2

   B. $Ca(HCO_3)_2$ and H_2O **D.** $CaCO_3$ and H_2O

                             E. $Ca(C_2H_3O_2)_2$ and H_2O

13. What is the conjugate base of water?

   A. OH^- *(aq)*   **B.** H^+ *(aq)*   **C.** H_2O *(l)*   **D.** H_3O^+ *(aq)*   **E.** O^{2-} *(aq)*

14. Identify the acid or base for each substance in the reaction $HSO_4^- + H_2O \rightleftharpoons OH^- + H_2SO_4$

   A. HSO_4^- acts as a base, H_2O acts as a base, OH^- acts as a base, H_2SO_4 acts as an acid

   B. HSO_4^- acts as an acid, H_2O acts as a base, OH^- acts as a base, H_2SO_4 acts as an acid

   C. HSO_4^- acts as a base, H_2O acts as an acid, OH^- acts as an acid, H_2SO_4 acts as a base

   D. HSO_4^- acts as an acid, H_2O acts as a base, OH^- acts as an acid, H_2SO_4 acts as a base

   E. HSO_4^- acts as a base, H_2O acts as an acid, OH^- acts as a base, H_2SO_4 acts as an acid

15. Which of the following could NOT be a Brønsted-Lowry acid?

   A. CN^- **B.** HS^- **C.** HF   **D.** $HC_2H_3O_2$ **E.** H_2SO_4

16. Which of the following indicators is yellow in an acidic solution and blue in a basic solution?

   A. methyl red   **C.** bromthymol blue

   B. phenolphthalein   **D.** all of the above

                             E. none of the above

17. Which of the following does NOT act as a Brønsted-Lowry acid?

   A. CO_3^{2-} **B.** HS^- **C.** HSO_4^- **D.** H_2O **E.** H_2SO_4

                                       Copyright © 2015 Sterling Test Prep

Questions **18-22** are based on the following titration graph.

Assume that the unknown acid is completely titrated with NaOH.

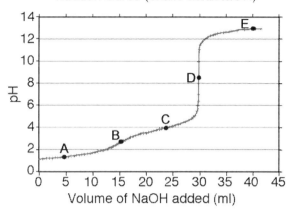

18. The unknown acid shown in the graph must be:

   A. monoprotic acid   **C.** triprotic acid

   B. diprotic acid   **D.** weak acid   **E.** none of the above

19. The pK_{a2} for this acid is located at point:

   A. A   **B.** B   **C.** C   **D.** D   **E.** E

20. At which point does the acid exist at 50% fully protonated and 50% singly deprotonated?

   A. A   **B.** B   **C.** C   **D.** D   **E.** E

21. At which point is the acid 100% singly deprotonated?

   A. A   **B.** B   **C.** C   **D.** D   **E.** E

22. Which points are the best buffer regions?

   A. A and B   **C.** B and D

   B. A and C   **D.** C and B   **E.** C and D

23. What is the term for a solution that is a good conductor of electricity?

   A. strong electrolyte   **C.** non-electrolyte

   B. weak electrolyte   **D.** aqueous electrolyte   **E.** none of the above

24. Which statement concerning the Arrhenius acid-base theory is NOT correct?

   A. Neutralization reactions produce H_2O plus a salt

   B. Acid-base reactions must take place in aqueous solution

   C. Arrhenius acids produce H^+ in H_2O solution

   D. Arrhenius bases produce OH^- in H_2O solution

   E. All are correct

25. Sodium hydroxide is a very strong base. If a concentrated solution of NaOH spills on a latex glove, it feels like water. Why is it that if the solution were to splash directly on a person's skin, it feels very slippery?

  A. As a liquid, NaOH is slippery, but this cannot be detected through a latex glove because of the friction between the latex surfaces
  B. NaOH destroys skin cells on contact and the remnants of skin cells feel slippery because the cells still contain natural oils
  C. NaOH lifts oil directly out of the skin cells and the extruded oil causes the slippery sensation
  D. NaOH reacts with skin oils, transforming them into soap
  E. NaOH, as a liquid, causes the skin to feel slippery from low viscosity

26. Which of the following pairs of substances could NOT function as a buffer system in aqueous solution?

  A. $NaHCO_3$ and Na_2CO_3
  B. HF and LiF
  C. $HClO_4$ and $KClO_4$
  D. HNO_2 and $NaNO_2$
  E. none could function as a buffer

27. Which of the following is an example of an Arrhenius acid?

  A. $HCl\,(aq)$
  B. $H_2SO_4\,(aq)$
  C. $HNO_3\,(aq)$
  D. None of the above
  E. All of the above

28. Which of the following properties is NOT characteristic of a base?

  A. Has a slippery feel
  B. Is neutralized by an acid
  C. Has a bitter taste
  D. Produces H^+ in water
  E. All of the above

29. In the expression $HCl\,(aq) \rightleftharpoons H^+ + Cl^-$, Cl is a:

  A. weak conjugate base
  B. strong conjugate base
  C. weak conjugate acid
  D. strong conjugate acid
  E. strong conjugate base and weak conjugate acid

30. What is the pH of an aqueous solution if the $[H^+] = 0.1$ M?

  A. 0   **B.** 1   **C.** 2   **D.** 10   **E.** 13

31. Which of the following is the strongest weak acid?

  A. H_2CO_3 $(Ka = 4.3 \times 10^{-7})$
  B. HNO_2 $(Ka = 4.5 \times 10^{-4})$
  C. HCN $(Ka = 4.9 \times 10^{-10})$
  D. HClO $(Ka = 3.0 \times 10^{-8})$
  E. HF $(Ka = 6.8 \times 10^{-4})$

32. Which of the following statements describes a neutral solution?

  A. $[H_3O^+] / [OH^-] = 1 \times 10^{-14}$
  B. $[H_3O^+] / [OH^-] = 1$
  C. $[H_3O^+] < [OH^-]$
  D. $[H_3O^+] > [OH^-]$
  E. $[H_3O^+] \times [OH^-] \neq 1 \times 10^{-14}$

    Copyright © 2015 Sterling Test Prep

33. Each of the following can act like a Brønsted-Lowry acid and base, EXCEPT:

    A. H_2O **B.** HCO_3^- **C.** $H_2PO_4^-$ **D.** NH_4^+ **E.** HS^-

34. If a light bulb in a conductivity apparatus glows brightly when testing a solution, which of the following must be true about the solution?

    A. It is highly reactive   **C.** It is highly ionized
    B. It is slightly reactive   **D.** It is slightly ionized
                                   E. It is not an electrolyte

35. What is the pH of a solution that has a $[H_3O^+] = 1.2 \times 10^{-3}$?

    A. 1.40   **B.** 2.92   **C.** 6.20   **D.** 8.80   **E.** 11.34

36. Which compound is both a Lewis base and a Brønsted-Lowry base?

    A. NH_3 **B.** NO_3 **C.** $Ca(OH)_2$ **D.** BH_3 **E.** NH_4^+

37. Which of the following explains why distilled H_2O is neutral?

    A. $[H^+] = [OH^-]$ **C.** Distilled H_2O has no H^+
    B. Distilled H_2O has no OH^- **D.** Distilled H_2O has no ions
                                   E. None of the above

38. If the concentration of H_3O^+ is 3.5×10^{-3} M, what is the molar concentration of OH^-?

    A. 3.5×10^{-10} **C.** 1.0×10^{-3}
    B. 1.0×10^{-7} **D.** 10.5×10^{-4} **E.** 2.9×10^{-12}

39. A strong acid tends to:

    A. form positively charged ions when dissolved in H_2O **C.** be strongly polar
    B. form negatively charged ions when dissolved in H_2O **D.** all of the above
                                                          E. none of the above

40. Which of the following is NOT a strong base?

    A. $Ca(OH)_2$ **B.** $Fe(OH)_3$ **C.** KOH **D.** $NaOH$ **E.** NH_2^-

41. What is the term for a substance that donates a proton in an acid-base reaction?

    A. Brønsted-Lowry acid   **C.** Arrhenius acid
    B. Brønsted-Lowry base   **D.** Arrhenius base   **E.** None of the above

42. In an acidic solution, the pH is:

    A. > 9 **B.** < 7 **C.** > 7 **D.** $= 7$ **E.** cannot be determined

43. Which is a strong acid and completely dissociates in H_2O?

    A. H_3PO_4 **B.** CH_3CH_2COOH **C.** HF **D.** HNO_3 **E.** H_2O

44. If a vinegar sample has a pH of 5, the solution is:

A. weakly basic

C. weakly acidic

B. neutral

D. strongly acidic

E. strongly basic

45. Which compound has a very large value of K_a in aqueous solution?

A. KOH

B. H_3PO_4

C. HNO_3

D. NaCl

E. NH_3

46. What is the best explanation for why a person might wash their hands with ashes?

A. The oils on the skin act as a base and react with the ashes to produce soap

B. After being burnt in the fire, the acids and bases of the log are neutralized, making it a gentle material to use on the hands

C. The ashes act as a base and react with skin oils to produce solutions of soap

D. The ashes act as an acid and react with the skin oils to produce soap

E. The amino acids on the skin act as a base and react with the ashes to produce soap

47. Which is incorrectly classified as an acid, a base or an amphoteric species?

A. LiOH / base

C. H_2S / acid

B. HS^- / amphoteric

D. NH_4^+ / base

E. none of the above

48. Which of the following is a general property of an acidic solution?

A. pH < 7

C. Tastes sour

B. Neutralizes bases

D. Turns litmus paper red

E. All of the above

49. What is the conjugate base of ^-OH?

A. O_2

B. O^{2-}

C. H_2O

D. O^-

E. H_3O^+

50. Which is the strongest acid?

Monoprotic Acids	Ka
Acid I	1×10^{-8}
Acid II	1×10^{-9}
Acid III	3.5×10^{-10}
Acid IV	2.5×10^{-8}

A. I

B. II

C. III

D. IV

E. Not enough data to determine

51. The $[^-OH]$ and the pH of 0.035 M KOH at 25 °C are, respectively:

A. 0.035 M and +12.5

C. 0.035 M and −1.46

B. 2.9×10^{-13} M and −12.5

D. 0.035 M and +1.46

E. 2.9×10^{-13} and +12.5

52. What is the conjugate acid of HS^-?

A. S^-

B. HS^-

C. H_2S

D. S^{2-}

E. S

Copyright © 2015 Sterling Test Prep

53. Which of the following is a weak acid?

    A. HCl (*aq*) (~100% ionized)

    B. $HC_2H_3O_2$ (*aq*) (~1% ionized)

    C. HNO_3 (*aq*) (~100% ionized)

    D. All of the above

    E. None of the above

54. What is the conjugate acid of water?

    A. H^+ (*aq*)   **B.** OH^- (*aq*)   **C.** O_2^- (*aq*)   **D.** H_2O (*l*)   **E.** H_3O^+ (*aq*)

55. In the reaction below, what does the symbol $\rightleftharpoons$ indicate?

$$^-OH + NH_4^+ \rightleftharpoons H_2O + NH_3$$

    A. The rate of the backward reaction is the same as the forward reaction; H^+ is not transferred

    B. The forward reaction does not progress

    C. The reaction cannot decide which direction produces equilibrium

    D. The forward and backward reactions are occurring simultaneously

    E. None of the above

56. Which of these salts is basic in an aqueous solution?

    A. NaF   **B.** $CrCl_3$ **C.** KBr   **D.** NH_4ClO_4 **E.** None are basic

57. What are the products from the neutralization reaction HNO_3 (*aq*) + $Ba(OH)_2$ (*aq*) $\rightarrow$?

    A. $Ba(NO_2)_2$ and H_2O

    B. $Ba(NO_3)_2$ and H_2O

    C. Ba_3N_2 and H_2O

    D. $Ba(NO_2)_2$ and H_2

    E. $Ba(NO_3)_2$ and H_2

58. At which pH will the net charge on an amino acid be zero, if the pI for it is 7.5?

    A. 2.5   **B.** 5.0   **C.** 7.0   **D.** 7.5   **E.** 10.0

59. Which molecule is a Lewis acid?

    A. NO_3^- **B.** NH_3 **C.** NH_4^+ **D.** CH_3COOH **E.** BH_3

60. A 36.0 mL sample of aqueous sulfuric acid was titrated with 0.250 M NaOH (*aq*) until neutralized. The residue was dried and then weighed with a mass of 861 mg. What is the molarity of the sulfuric acid solution?

    A. 0.0732 M   **B.** 0.168 M   **C.** 0.262 M   **D.** 0.445 M   **E.** 0.894 M

61. What is the pH of an aqueous solution if $[H^+] = 0.000001$ M?

    A. 1   **B.** 4   **C.** 6   **D.** 7   **E.** 9

62. The classification as weak or strong for an acid or base is determined by:

    A. the concentration of the acid or base

    B. the extent of dissociation of the dissolved acid or base

    C. the solubility of the acid or base

    D. the ability to be neutralized by buffering solution

    E. the more than one choice is correct

63. Which of the following statements about strong or weak acids is true?

    A. A weak acid reacts with a strong base

    B. A strong acid does not react with a strong base

    C. A weak acid readily forms ions when dissolved in water

    D. A weak acid and a strong acid at the same concentration are equally corrosive

    E. A strong base does not react with a strong acid

64. Which is NOT correctly classified as a base, salt or an amphoteric species?

    A. KF / salt   **C.** HCO_3^- / amphoteric

    B. S^{2-} / amphoteric   **D.** NH_3 / base   **E.** all are correctly classified

65. What is the term for a substance that releases H^+ in H_2O?

    A. Brønsted-Lowry acid   **C.** Arrhenius acid

    B. Brønsted-Lowry base   **D.** Arrhenius base   **E.** none of the above

66. Which compounds forms CO_2 and H_2O when reacted with an acid?

    A. $NaC_2H_3O_2$ **B.** NH_3 **C.** $CaCO_3$ **D.** $Mg(OH)_2$ **E.** none of the above

67. The conjugate bases of HSO_4^-, CH_3OH and H_3O^+ are, respectively:

    A. SO_4^{2-}, CH_2OH^- and ^-OH **C.** SO_4^-, CH_3O^- and ^-OH

    B. CH_3O^-, SO_4^{2-} and H_2O **D.** SO_4^-, CH_2OH^- and H_2O

                                       E. SO_4^{2-}, CH_3O^- and H_2O

68. What is the term used for an H^+ acceptor?

    A. proton acceptor   **C.** Arrhenius acid

    B. Brønsted-Lowry acid   **D.** amphiprotic   **E.** none of the above

69. An acid is represented by which of the following?

    A. HBr   **B.** C_2H_6 **C.** KOH   **D.** NaF   **E.** $NaNH_2$

70. Which of the following statements describes an acidic solution?

    A. $[H_3O^+] / [^-OH] = 1 \times 10^{-14}$ **C.** $[H_3O^+] < [^-OH]$

    B. $[H_3O^+] \times [^-OH] \neq 1 \times 10^{-14}$ **D.** $[H_3O^+] > [^-OH]$ **E.** $[H_3O^+] / [^-OH] = 1$

71. According to the Brønsted-Lowry theory, a base is a:

    A. hydroxide ion acceptor   **C.** proton donor

    B. proton acceptor   **D.** hydronium ion donor

                                     E. lone pair acceptor

72. In the following reaction, which reactant is a Brønsted-Lowry base?

$$H_2CO_3\,(aq) + Na_2HPO_4\,(aq) \rightarrow NaHCO_3\,(aq) + NaH_2PO_4\,(aq)$$

    A. $NaHCO_3$ **C.** Na_2HPO_4

    B. NaH_2PO_4 **D.** H_2CO_3 **E.** none of the above

                                                            Copyright © 2015 Sterling Test Prep

73. Complete neutralization of phosphoric acid with barium hydroxide yields $Ba_3(PO_4)_2$ as one of the products when separated and dried. This indicates that phosphoric acid is a:

A. monoprotic acid

B. diprotic acid

C. hexaprotic acid

D. tetraprotic acid

E. triprotic acid

74. The isoelectric point of an amino acid is defined as the:

A. pH equals the pKa

B. pH at which the amino acid exists in the acidic form

C. pH at which the amino acid exists in the basic form

D. pH at which the amino acid exits in the zwitterion form

E. pH at which the amino acid exists in the protonated form

75. When 37 g of $Ca(OH)_2$ is added to 1000 ml of 0.55 M H_2SO_4, the resulting pOH is:

A. 3   B. 8   C. 11   D. 13   E. 14

76. Which reactant is a Brønsted-Lowry acid in HCl (*aq*) + KHS (*aq*) → KCl (*aq*) + H_2S (*aq*)?

A. KCl   B. H_2S C. HCl   D. KHS   E. None of the above

77. Citric acid is a triprotic acid with three carboxylic acid groups having pK_a values of 3.1, 4.8 and 6.4. At a pH of 5.5, what is the predominant protonation state of citric acid?

A. All 3 carboxylic acid groups are deprotonated

B. All 3 carboxylic acid groups are protonated

C. 1 carboxylic acid group is deprotonated while 2 are protonated

D. 2 carboxylic acid groups are deprotonated while 1 is protonated

E. The protonation state cannot be determined

78. Which of the following is the strongest base?

A. ClO_3^- B. NH_3 C. ClO^- D. ClO_2^- E. ClO_4^-

79. Which of the following compound–classification pair is incorrectly matched?

A. HF – weak acid

B. $LiC_2H_3O_2$ – salt

C. NH_3 – weak base

D. HI – strong acid

E. $Ca(OH)_2$ – weak base

80. Which of the following indicators is orange at pH 5?

A. phenolphthalein

B. methyl red

C. bromthymol blue

D. all of the above

E. none of the above

81. Which of the following is a strong base?

A. $Ba(OH)_2$ B. CH_3COOH C. NH_3 D. CH_3OH E. NaCl

82. Which one is a Brønsted-Lowry acid?

A. NH_3 B. NO_3^- C. CH_3COOH D. BH_3 E. CH_2Cl_2

83. The chemical species present in all acidic solutions is:

    A. H_2O^+ (*aq*)   **B.** H_3O^+ (*l*)   **C.** H_2O (*aq*)   **D.** ^-OH (*aq*)   **E.** H_3O^+ (*aq*)

84. Which of the following is a general property of an acidic solution?

    A. pH greater than 7   **C.** Feels slippery

    B. Turns litmus paper blue   **D.** Tastes sour   **E.** None of the above

85. The H_3O^+ ion is called the:

    A. protium ion   **C.** hydronium ion

    B. hydrogen ion   **D.** hydroxide ion   **E.** water ion

86. Which molecule is acting as a base in the following reaction $^-OH + NH_4^+ \rightarrow H_2O + NH_3$?

    A. ^-OH **B.** NH_4^+ **C.** H_2O **D.** NH_3 **E.** H_3O^+

87. Which of the following is NOT a conjugate acid/base pair?

    A. S^{2-} / H_2S **C.** $H_2O / ^-OH$

    B. HSO_4^- / SO_4^{2-} **D.** PH_4^+ / PH_3 **E.** all are conjugate acid/base pairs

88. In aqueous solution, what is the term for ions that do not participate in a reaction and do not appear in the net ionic equation?

    A. zwitterions   **C.** nonelectrolyte ions

    B. spectator ions   **D.** electrolyte ions   **E.** none of the above

89. An Arrhenius acid is defined as a substance that:

    A. decreases $[H^+]$ when placed in aqueous solutions

    B. increases $[H^+]$ when placed in aqueous solutions

    C. acts as a proton acceptor in any system

    D. acts as a proton donor in any system

    E. acts as a proton acceptor in aqueous solutions

90. If [HF] (*aq*), $[H^+]$ and $[F^-]$ at equilibrium are 2.5×10^{-1} M, 5.0×10^{-2} M, and 5.0×10^{-5} M respectively, what is the dissociation constant for HF(*aq*) in the reaction $HF \ (aq) \rightleftharpoons H^+ + F^-$?

    A. 1×10^{-5} **C.** 3.0×10^{-3}

    B. 2.5×10^{-4} **D.** 5.0×10^{-5} **E.** 3.0×10^{-2}

91. Given the pK_a values for phosphoric acid of 2.14, 6.86 and 12.4, what is the ratio of $HPO_4^{2-} / H_2PO_4^-$ in a typical muscle cell when the pH is 7.2?

    A. 6.3×10^{-6} **B.** 1.1×10^5 **C.** 0.46   **D.** 2.2   **E.** 3.1×10^3

92. What is the sum of the coefficients for the balanced molecular equation?

    $BaCO_3 \ (aq) + HNO_3 \ (aq) \rightarrow \ ?$

    A. 9   **B.** 10   **C.** 8   **D.** 3   **E.** 6

Copyright © 2015 Sterling Test Prep

93. What is the term for a substance that changes color according to the pH of the solution?

   A. Arrhenius acid   **C.** Acid-base indicator

   B. Brønsted-Lowry acid   **D.** Acid-base signal   **E.** None of the above

94. Which set below contains *only* weak electrolytes?

   A. NH_4Cl (*aq*), $HClO_2$ (*aq*), HCN (*aq*)   **C.** KOH (*aq*), H_3PO_4 (*aq*), $NaClO_4$ (*aq*)

   B. NH_3 (*aq*), $HC_2H_3O_2$ (*aq*), HCN (*aq*)   **D.** HNO_3 (*aq*), H_2SO_4 (*aq*), HCN (*aq*)

                                            E. NaOH (*aq*), H_2SO_4 (*aq*), $HC_2H_3O_2$ (*aq*)

95. Compared to a solution with a higher pH, a solution with a lower pH has a(n):

   A. decreased Ka   **C.** increased pK_a

   B. increased $[^-OH]$ **D.** increased $[H^+]$ **E.** decreased $[H^+]$

96. Which is the correct net ionic equation for the hydrolysis reaction of Na_2S?

   A. Na^+ (*aq*) + H_2O (*l*) → NaOH (*aq*) + H_2 (*g*)

   B. Na^+ (*aq*) + $2H_2O$ (*l*) → NaOH (*aq*) + H_2O^+ (*aq*)

   C. S^{2-} (*aq*) + H_2O (*l*) → $2HS^-$ (*aq*) + ^-OH (*aq*)

   D. S^{2-} (*aq*) + $2H_2O$ (*l*) → HS^- (*aq*) + H_3O^+ (*aq*)

   E. S^{2-} (*aq*) + H_2O (*l*) → HS^- (*aq*) + ^-OH (*aq*)

97. What is the molarity of a hydrochloric acid solution if 20.00 mL of HCl is required to neutralize 0.424 g of sodium carbonate (105.99 g/mol)?

$$2HCl\ (aq) + Na_2CO_3\ (aq) \rightarrow 2NaCl\ (aq) + H_2O\ (l) + CO_2\ (g)$$

   A. 0.150 M   **B.** 0.250 M   **C.** 0.300 M   **D.** 0.400 M   **E.** 0.500 M

98. What is the $[H_3O^+]$ of a solution that has a pH = 2.34?

   A. 1.3×10^1 M   **C.** 4.6×10^{-3} M

   B. 2.3×10^{-10} M   **D.** 2.4×10^{-3} M   **E.** $3.6\ 10^{-8}$ M

99. Which compound is amphoteric?

   A. NO_3 **C.** $CH_3CH_2COO^-$

   B. HBr   **D.** BH_3 **E.** HSO_4

100. If 10.0 mL of 0.100 M HCl is titrated with 0.200 M NaOH, what volume of sodium hydroxide solution is required to neutralize the acid?

$$HCl\ (aq) + NaOH\ (aq) \rightarrow NaCl\ (aq) + H_2O\ (l)$$

   A. 5.00 mL   **B.** 10.0 mL   **C.** 20.0 mL   **D.** 40.0 mL   **E.** 80.0 mL

101. A requirement for a Brønsted-Lowry base is the:

   A. production of H_3O^+ upon reaction with H_2O **C.** presence of ^-OH in its formula

   B. lone pair of electrons in its Lewis dot structure   **D.** presence of H_2O as a reaction medium

                                                    E. presence of a metal ion in its formula

Copyright © 2015 Sterling Test Prep

102. Can an acid and a base react to form an acid?

   A. The reaction can only occur if the reactants are organic/carboxylic acids and organic bases

   B. An acid and a base react to form an acid if the acid is a weak acid and the base is a strong base

   C. No, acids always form salts when reacted with bases

   D. The reaction can only occur if the substances reacting are true acids and bases

   E. An acid and a base react to form an acid if the acid is a strong acid and the base is a weak base

103. The Brønsted-Lowry acid and base in $NH_4^+ + CN^- \rightarrow NH_3 + HCN$ are, respectively:

   A. NH_4^+ and CN^- **C.** NH_4^+ and HCN

   B. CN^- and HCN   **D.** NH_3 and CN^- **E.** NH_3 and NH_4^+

104. In the following reaction, which reactant is a Brønsted-Lowry acid?

$$H_2CO_3\,(aq) + Na_2HPO_4\,(aq) \rightarrow NaHCO_3\,(aq) + NaH_2PO_4\,(aq)$$

   A. NaH_2PO_4 **C.** H_2CO_3

   B. $NaHCO_3$ **D.** Na_2HPO_4 **E.** None of the above

105. Which one of the following is the weakest acid?

   A. HF   ($K_a = 6.5 \times 10^{-4}$)   **C.** HClO   ($K_a = 3.0 \times 10^{-8}$)

   B. HNO_2 ($K_a = 4.5 \times 10^{-4}$)   **D.** HCN   ($K_a = 6.3 \times 10^{-10}$)

                                       E. HI   ($K_a = 7.1 \times 10^{2}$)

106. Acids and bases react to form:

   A. Brønsted-Lowry acids   **C.** Lewis acids

   B. Arrhenius acids   **D.** Lewis bases   **E.** salts

107. Which of the following is a general property of a basic solution?

   A. Turns litmus paper red   **C.** Tastes sour

   B. pH less than 7   **D.** Feels slippery   **E.** None of the above

108. A weak acid is titrated with a strong base. When the concentration of the conjugate base is equal to the concentration of the acid, the titration is at the:

   A. end point   **B.** buffering region

   C. equivalence point   **D.** diprotic point   **E.** indicator zone

109. Which of the following statements is NOT true about a neutralization reaction?

   A. Water is always formed in a neutralization reaction

   B. A neutralization is the reaction of an ^-OH with an H^+ **D.** All of the above are true

   C. One molecule of acid neutralizes one molecule of base   **E.** None of the above is true

110. Which of the following is NOT a strong acid?

   A. HBr *(aq)*   **B.** HNO_3 **C.** H_2CO_3 **D.** H_2SO_4 **E.** HCl *(aq)*

 Copyright © 2015 Sterling Test Prep

111. What is the color of phenolphthalein indicator at pH 7?

    A. pink   **B.** colorless   **C.** red   **D.** blue   **E.** none of the above

112. A Brønsted-Lowry acid is defined as a substance that:

    A. acts as a H^+ acceptor in any system   **C.** increases $[H^+]$ when placed in H_2O

    B. acts as a proton donor in any system   **D.** decreases $[H^+]$ when placed in H_2O

                                       E. acts as lone pair acceptor in any system

113. Which acts as the best buffer solution:

    A. Strong acids or bases   **C.** Salts

    B. Strong acids and their salts   **D.** Weak acids or bases and their salts

                                       E. Strong bases and their salts

114. When fully neutralized by treatment with barium hydroxide, a phosphoric acid yields $Ba_2P_2O_7$ as one of its products. The parent acid for the anion in this compound is:

    A. monoprotic acid   **C.** triprotic acid

    B. diprotic acid   **D.** hexaprotic acid   **E.** tetraprotic acid

115. What is the term used as a synonym for hydrogen ion donor?

    A. Brønsted-Lowry base   **C.** amphiprotic

    B. proton donor   **D.** Arrhenius base   **E.** none of the above

116. Which compound has a value of K_a that is close to 10^{-5}?

    A. $CH_3CH_2CO_2H$ **B.** KOH   **C.** NaCl   **D.** HNO_3 **E.** NH_3

117. A base is anything that:

    A. can be used to clean drains   **C.** accepts an ^-OH

    B. donates an ^-OH **D.** accepts an H^+ **E.** has a bitter taste

118. In which of the following pairs of acids are both chemical species in the pair weak acids?

    A. $HC_2H_3O_2$ and HI   **C.** HCN and H_2S

    B. H_2CO_3 and HBr   **D.** H_3PO_4 and H_2SO_4 **E.** HCl and HBr

119. Which of the following indicators is colorless in an acidic solution and pink in a basic solution?

    A. phenolphthalein   **C.** methyl red

    B. bromthymol blue   **D.** all of the above   **E.** none of the above

120. Which of the following is NOT a strong acid?

    A. HI (*aq*)   **B.** $HClO_4$ **C.** HCl (*aq*)   **D.** HNO_3 **E.** $HC_2H_3O_2$

121. What is the name given to a solution that resists changes in pH?

    A. neutral   **B.** basic   **C.** buffer   **D.** protic   **E.** acidic

122. Which would NOT be used to make a buffer solution?

    A. H_2SO_4 **B.** H_2CO_3 **C.** NH_4OH **D.** CH_3COOH **E.** Tricene

123. In the following reaction, which reactant is a Brønsted-Lowry base?

    $HCl\ (aq) + KHS\ (aq) \rightarrow KCl\ (aq) + H_2S\ (aq)$

    A. H_2S **B.** KCl   **C.** KHS   **D.** HCl   **E.** none of the above

124. The ions, Ca^{2+}, Mg^{2+}, Fe^{2+}/Fe^{3+}, which are present in all ground water, can be removed by pretreating the water with:

    A. $PbSO_4$ **C.** KNO_3

    B. $Na_2CO_3 \cdot 10H_2O$ **D.** $CaCl_2$ **E.** 0.05 M HCl

125. What is the pI for glutamic acid that contains two carboxylic acid groups (pK_a 2.2 and 4.2) and an amino group (pK_a 9.7)?

    A. 3.2   **B.** 1.0   **C.** 6.4   **D.** 5.4   **E.** 5.95

126. Which of the following is a diprotic acid?

    A. HCl   **B.** H_3PO_4 **C.** HNO_3 **D.** H_2SO_3 **E.** H_2O

127. What are the products from the complete neutralization of carbonic acid with aqueous potassium hydroxide?

    A. $KHCO_4\ (aq)$ and $H_2O\ (l)$ **C.** $K_2CO_3\ (aq)$ and $H_2O\ (l)$

    B. $KC_2H_3O_2\ (aq)$ and $H_2O\ (l)$ **D.** $KHCO_3\ (aq)$ and $H_2O\ (l)$

                                         E. $K_2C_2H_3O_2\ (aq)$ and $H_2O\ (l)$

128. Calculate the pH of 0.00756 M HNO_3.

    A. 2.1   **B.** 12.9   **C.** 11.7   **D.** 7. 93   **E.** 5.67

129. Salts of strong acids and strong bases are:

    A. neutral   **B.** basic   **C.** acidic   **D.** salts   **E.** none of the above

130. Which of the following is a general property of a basic solution?

    A. Neutralizes acids   **C.** Feels slippery

    B. Turns litmus paper blue   **D.** Tastes bitter   **E.** All of the above

131. At 25 °C, the value of K_w is:

    A. 1.0   **B.** 1.0×10^{-7} **C.** 1.0×10^{-14} **D.** 1.0×10^{7} **E.** 1.0×10^{14}

132. The main component of bleach is sodium hypochlorite ($NaOCl$) which consists of sodium ions, Na^+ and hypochlorite ions (^-OCl). What products are formed when this compound is reacted with the hydrochloric acid (HCl) present in toilet bowl cleaner?

    A. $NaOH$, O_2 and Cl_2 **C.** $NaOH$, H_2O and Cl_2

    B. $NaCl$ and $HOCl$ **D.** $NaCl$, O_2 and $HClO_2$ **E.** $NaCl$ and $NaOH$

 Copyright © 2015 Sterling Test Prep

133. Which of the following pairs of chemical species contains two polyprotic acids?

   A. HNO_3 and $H_2C_4H_4O_6$ **C.** H_3PO_4 and HCN

   B. $HC_2H_3O_2$ and $H_3C_6H_5O_7$ **D.** H_2S and H_2CO_3 **E.** HCN and HNO_3

134. Which of the following is a general property of an acidic solution?

   A. Turns litmus paper blue   **C.** Tastes bitter

   B. Neutralizes acids   **D.** Feels slippery   **E.** None of the above

135. Which one of the following is the strongest weak acid?

   A. CH_3COOH ($K_a = 1.8 \times 10^{-5}$)   **C.** HCN   ($K_a = 6.3 \times 10^{-10}$)

   B. HF   ($K_a = 6.5 \times 10^{-4}$)   **D.** HClO   ($K_a = 3.0 \times 10^{-8}$)

                                          E. HNO_2 ($K_a = 4.5 \times 10^{-4}$)

136. Which acts as a buffer system?

   A. $NH_3 + H_2O \rightleftharpoons OH^- + NH_4^+$ **C.** $HC_2H_3O_2 \rightleftharpoons H^+ + C_2H_3O_2^-$

   B. $H_2PO_4 \rightleftharpoons H^+ + HPO_4^{2-}$ **D.** $CO_2 + H_2O \rightleftharpoons H_2CO_3 \rightleftharpoons HCO_3^- + H^+$

                                          E. All of the above

137. What is the term for a solution for which concentration has been established accurately, usually to three or four significant digits?

   A. standard solution   **C.** normal solution

   B. stock solution   **D.** reference solution   **E.** none of the above

138. The reaction most likely to lead to the formation of calcium bicarbonate in limestone regions is the reaction between:

   A. sodium hydroxide and acetic acid

   B. calcium carbonate and carbonic acid

   C. calcium chloride and sodium carbonate

   D. sodium nitrate and carbonic acid

   E. sodium carbonate and calcium carbonate

139. Which of the following statements is correct?

   A. In a basic solution, $[H_3O^+] > 10^{-7}$ and $[OH^-] < 10^{-7}$

   B. In a basic solution, $[H_3O^+] > 10^{-7}$ and $[OH^-] > 10^{-7}$

   C. In a basic solution, $[H_3O^+] < 10^{-7}$ and $[OH^-] < 10^{-7}$

   D. In a basic solution, $[H_3O^+] < 10^{-7}$ and $[OH^-] > 10^{-7}$

   E. In a basic solution, $[H_3O^+] > 10^{-7}$ and $[OH^-] = 10^{-7}$

140. The hydrogen sulfate ion HSO_4^- is amphoteric. In which of the following equations does it act as an acid?

   A. $HSO_4^- + {}^-OH \rightarrow H_2SO_4 + O^{2-}$ **C.** $HSO_4^- + H_3O^+ \rightarrow SO_3 + 2H_2O$

   B. $HSO_4^- + H_2O \rightarrow SO_4^{2-} + H_3O^+$ **D.** $HSO_4^- + H_2O \rightarrow H_2SO_4 + {}^-OH$

                                          E. None of the above

141. Which of the following is a general property of a basic solution?

   A. Neutralizes bases   **C.** Turns litmus paper red

   B. pH less than 7   **D.** Tastes sour   **E.** None of the above

142. In the following equation, which is the proton donor and which is the proton acceptor?

$$CO_3^{2-} (aq) + H_2O (l) \rightarrow HCO_3^- (aq) + OH^- (aq)$$

   A. OH^- is the donor and HCO_3^- is the acceptor   **C.** H_2O is the donor and CO_3^{2-} is the acceptor

   B. HCO_3^- is the donor and OH^- is the acceptor   **D.** CO_3^{2-} is the donor and H_2O is the acceptor

                                                 E. CO_3^{2-} is the donor and OH^- is the acceptor

143. What is the pH of 0.001 M HCl?

   A. 2   **B.** 1   **C.** 4   **D.** 3   **E.** 5

144. What are the products from the complete neutralization of sulfuric acid with aqueous sodium hydroxide?

   A. $Na_2SO_3 (aq)$ and $H_2O (l)$ **C.** $NaHSO_3 (aq)$ and $H_2O (l)$

   B. $NaHSO_4 (aq)$ and $H_2O (l)$ **D.** $Na_2S (aq)$ and $H_2O (l)$

                                         E. $Na_2SO_4 (aq)$ and $H_2O (l)$

145. What is the conjugate acid-base pair in the reaction $CH_3NH_2 + HCl \leftrightarrow CH_3NH_3^+ + Cl^-$?

   A. HCl and Cl^- **C.** CH_3NH_2 and Cl^-

   B. $CH_3NH_3^+$ and Cl^- **D.** CH_3NH_2 and HCl   **E.** HCl and H_3O^+

146. Does a solution become more or less acidic when a weak acid is added to a concentrated solution of HCl?

   A. Less acidic, because the concentration of OH^- increases

   B. No change in acidity, because [HCl] is too high to be changed by the weak solution

   C. Less acidic, because the solution becomes more dilute with a less concentrated solution of H_3O^+ being added to the solution

   D. More acidic, because more H_3O^+ are being added to the solution

   E. More acidic, because the solution becomes more dilute with a less concentrated solution of H_3O^+ being added to the solution

147. The neutralization of $Cr(OH)_3$ with H_2SO_4 produces which product?

   A. $Cr_2(SO_4)_3$ **C.** ^-OH

   B. SO_2 **D.** H_3O^+ **E.** H_2SO_4

148. In the following reaction, which reactant is a Brønsted-Lowry acid?

$$NaHS (aq) + HCN (aq) \rightarrow NaCN (aq) + H_2S (aq)$$

   A. NaCN   **B.** H_2S **C.** NaHS   **D.** HCN   **E.** none of the above

149. Relative to a pH of 7, a solution with a pH of 4 has:

   A. 30 times less $[H^+]$ **C.** 1000 times greater $[H^+]$

   B. 300 times less $[H^+]$ **D.** 300 times greater $[H^+]$ **E.** 30 times greater $[H^+$

 Copyright © 2015 Sterling Test Prep

150. If a salt to acid ratio is 1:10 for an acid with $K_a = 1 \times 10^{-4}$, what is the pH of the solution?

    A. 2   **B.** 3   **C.** 4   **D.** 5   **E.** 6

151. What is the term for a substance that releases hydroxide ions in water?

    A. Brønsted-Lowry base   **C.** Arrhenius base

    B. Brønsted-Lowry acid   **D.** Arrhenius acid

                                        E. None of the above

152. Calculate the hydrogen ion concentration in a solution with a pH = 6.35.

    A. 4.5×10^{-7} M   **C.** 7.65 M

    B. 7.55×10^{-8} M   **D.** 6.35 M

                                        E. 6.35×10^{-8} M

153. Which of the following would be predominantly deprotonated in the stomach if gastric juice has a pH of about 2?

    A. phosphoric acid ($pK_a = 2.2$)   **C.** acetic acid ($pK_a = 4.7$)

    B. lactic acid, ($pK_a = 3.9$)   **D.** phenol ($pK_a = 9.8$)

                                        E. hydrochloric acid ($pK_a = -6$)

154. A metal and a salt solution react only if the metal introduced into the solution is:

    A. below the replaced metal in the activity series

    B. above the replaced metal in the activity series

    C. below hydrogen in the activity series

    D. above hydrogen in the activity series

    E. equal to the replaced metal in the activity series

155. What volume of barium hydroxide is required to neutralize the acid when 25.0 mL of 0.100 M HCl is titrated with 0.150 M $Ba(OH)_2$?

$$2 \text{ HCl } (aq) + Ba(OH)_2 \, (aq) \rightarrow BaCl_2 \, (aq) + 2 \text{ H}_2\text{O } (l)$$

    A. 32.4 mL   **B.** 25.0 mL   **C.** 16.7 mL   **D.** 8.33 mL   **E.** 37.5 mL

156. Identify the Brønsted-Lowry acid and base, respectively: $NH_3 + HCN \rightarrow NH_4^+ + CN^-$

    A. NH_3 and NH_4^+ **C.** HCN and NH_3

    B. NH_4^+ and CN^- **D.** NH_3 and HCN   **E.** NH_3 and NH_4^+

157. Which of the following is NOT an example of an acidic salt?

    A. sodium hydrogen sulfate   **C.** barium dihydrogen phosphate

    B. nickel (II) bichromate   **D.** aluminum bicarbonate

                                        E. potassium hydrogen chloride

158. Calculate $[F^-]$ in a 2 M solution of hydrogen fluoride if the K_a of HF is 6.5×10^{-4}.

    A. 3.6×10^{-2} M   **C.** 1.9×10^{-2} M

    B. 1.7×10^{-2} M   **D.** 1.3×10^{-3} M   **E.** 6.5×10^{-2} M

159. Which of the following must be true if an unknown solution is a poor conductor of electricity?

A. Solution is slightly reactive

B. Solution is highly corrosive

C. Solution is highly ionized

D. Solution is slightly ionized

E. Solution is highly reactive

160. What is the pH of a buffer solution, where acetic acid (with a K_a of 1.8×10^{-5}) and its conjugate base are in a 10:1 ratio?

A. 1.9   B. 3.7   C. 5.7   D. 7.0   E. 7.4

161. Which characteristic of a molecule describes an acid?

A. Donates hydrogen atoms

B. Dissolves metal

C. Donates hydrogen ions

D. Accepts hydrogen atoms

E. Donates hydronium ions

162. Which of the following pairs of acids and conjugate bases is NOT correctly labeled?

Acid	Conjugate Base
A. NH_4^+	NH_3
B. HSO_3^-	SO_3^{2-}
C. H_2SO_4	HSO_4^-
D. HSO_4^-	SO_4^{2-}
E. HFO_2	HFO_3

163. What is the pH of an aqueous solution if the $[H^+] = 0.001$ M?

A. 1   B. 2   C. 3   D. 10   E. 11

164. Which of the following is NOT a strong base?

A. $Ca(OH)_2$ B. KOH   C. NaOH   D. $Al(OH)_3$ E. ^-OH

165. What volume of 0.15 M H_2SO_4 is needed to neutralize 40 ml of 0.2 M NaOH?

A. 27 ml   B. 40 ml   C. 65 ml   D. 80 ml   E. 105 ml

166. In the following titration curve, what does the inflection point represent?

A. The weak acid is 50% protonated and 50% deprotonated

B. The pH where the solution functions most effectively as a buffer

C. Equal concentration of weak acid and conjugate base

D. pH of solution equals pK_a of weak acid

E. All of the above

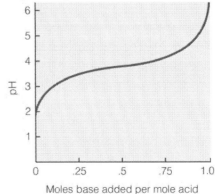

Copyright © 2015 Sterling Test Prep

167. What is the term for a substance that accepts a proton in an acid-base reaction?

    A. Brønsted-Lowry base   **C.** Arrhenius base

    B. Brønsted-Lowry acid   **D.** Arrhenius acid   **E.** None of the above

168. If $[H_3O^+]$ in an aqueous solution is 7.5×10^{-9} M, the $[OH^-]$ is:

    A. 6.4×10^{-5} M   **C.** 7.5×10^{-23} M

    B. $3.8 \times 10^{+8}$ M   **D.** 1.5×10^{-6} M   **E.** 9.0×10^{-9} M

169. Identify the acid/base behavior of each substance for $H_3O^+ + Cl^- \rightleftharpoons H_2O + HCl$

    A. H_3O^+ acts as an acid, Cl^- acts as a base, H_2O acts a base, HCl acts as an acid

    B. H_3O^+ acts as a base, Cl- acts as an acid, H_2O acts a base, HCl acts as an acid

    C. H_3O^+ acts as an acid, Cl^- acts as a base, H_2O acts an acid, HCl acts as a base

    D. H_3O^+ acts as a base, Cl^- acts as an acid, H_2O acts an acid, HCl acts as a base

    E. H_3O^+ acts as an acid, Cl^- acts as a base, H_2O acts a base, HCl acts as a base

170. Which of the following is a triprotic acid?

    A. HNO_3 **B.** H_3PO_4 **C.** H_2SO_3 **D.** $HC_2H_3O_2$ **E.** CH_2COOH

171. Which of the following is a strong acid?

    A. HNO_3 (*aq*) (~100% ionized)   **C.** HCl (*aq*) (~100% ionized)

    B. H_2SO_4 (*aq*) (~100% ionized)   **D.** All of the above

                                         E. None of the above

172. The K_a of a buffer is 4.5×10^{-4}. If the concentration of an undissociated weak acid is equal to the concentration of the conjugate base, the pH of this buffer system is between:

    A. 1 and 2   **B.** 3 and 4   **C.** 5 and 6   **D.** 7 and 8   **E.** 9 and 10

173. Which is the balanced equation for the neutralization reaction between $Al(OH)_3$ and HCl (*aq*)?

    A. $Al(OH)_3 + 3HCl$ (*aq*) $\rightarrow AlCl_3 + H^+ + OH^-$

    B. $Al^{3+} + OH^- + H^+ + Cl^- \rightarrow AlCl_3 + H_2O$

    C. $Al(OH)_3 + 3HCl$ (*aq*) $\rightarrow AlCl_3 + 3H_2O$

    D. $Al(OH)_3 + HCl$ (*aq*) $\rightarrow AlCl_3 + H_2O$

    E. $Al(OH)_3 + HCl$ (*aq*) $\rightarrow AlCl_3 + 3H_2O$

174. Which of the following is an example of an Arrhenius acid?

    A. $KC_2H_3O_2$ (*aq*)   **C.** $NH_4C_2H_3O_2$ (*aq*)

    B. $HC_2H_3O_2$ (*aq*)   **D.** All of the above   **E.** None of the above

175. When acids and bases react, the product other than water is a:

    A. hydronium ion   **C.** hydrogen ion

    B. metal   **D.** hydroxide ion   **E.** salt

176. What happens to the corrosive properties of an acid and a base after they neutralize each other?

   A. The corrosive properties are doubled because the acid and base are combined in the salt

   B. The corrosive properties remain the same when the salt is mixed into water

   C. The corrosive properties are neutralized because the acid and base no longer exist

   D. The corrosive properties are unaffected because salt is a corrosive agent

   E. The corrosive properties are increased because salt is a corrosive agent

177. Which of the following statements about the reaction of acids with metals is NOT correct?

   A. Metal atoms, when dissolved in acid, become positive metal ions

   B. Hydrogen gas is produced when a metal dissolves in an acid

   C. Any metal above hydrogen in the activity series dissolves in a non-oxidizing acid

   D. Acids react with many, but not all, metals

   E. Any metal below hydrogen in the activity series dissolves in a non-oxidizing acid

178. If an unknown solution is a good conductor of electricity, which of the following must be true?

   A. The solution is highly reactive

   B. The solution is slightly reactive

   C. The solution is highly ionized

   D. The solution is slightly ionized

   E. None of the above

179. Which of the following is the weakest acid?

   A. HCO_3^- $pK_a = 10.32$

   B. $HC_2H_3O_2$ $pK_a = 4.76$

   C. NH_4^+ $pK_a = 9.20$

   D. $H_2PO_4^-$ $pK_a = 7.18$

   E. CCl_3COOH $pK_a = 2.86$

180. What volume of 0.03 NaOH is needed to titrate 40 ml of 0.1 N H_3PO_4?

   A. 44ml   **B.** 88 ml   **C.** 133 ml   **D.** 266 ml   **E.** 399 ml

181. If stomach digestive juice has a pH of 2, the solution is:

   A. strongly acidic

   B. weakly basic

   C. neutral

   D. weakly acidic

   E. buffered

182. What is the $[H_3O^+]$ in a solution with pH = 11.61?

   A. 1.05×10^{-14} M

   B. 4.35×10^{-4} M

   C. 2.40×10^1 M

   D. 1.65×10^1 M

   E. 2.45×10^{-12} M

183. Which of the following statements describes a basic solution?

   A. $[H_3O^+] \times [OH^-] \neq 1 \times 10^{-14}$

   B. $[H_3O^+] / [OH^-] = 1 \times 10^{-14}$

   C. $[H_3O^+] > [OH^-]$

   D. $[H_3O^+] < [OH^-]$

   E. $[H_3O^+] / [OH^-] = 1$

184. What is the sum of the coefficients for the balanced molecular equation for the following acid-base reaction LiOH (*aq*) + H_2SO_4 (*aq*) → ?

   A. 6   **B.** 7   **C.** 8   **D.** 10   **E.** 12

Copyright © 2015 Sterling Test Prep

185. Which of the following indicators is pink at a pH of 9?

    A. methyl red   **C.** bromthymol blue

    B. phenolphthalein   **D.** all of the above   **E.** none of the above

186. In an acidic solution, $[H_3O^+]$ is:

    A. $< 1 \times 10^{-7}$ M   **C.** $< 1 \times 10^{-7}$ M

    B. $> 1 \times 10^{-7}$ M   **D.** 1×10^{-7} M   **E.** cannot be determined

187. Since $pK_a = -\log K_a$, which of the following is a correct statement?

    A. Since the pK_a for conversion of the ammonium ion to ammonia is 9.25, ammonia is a weaker base then the acetate ion

    B. For carbonic acid with pK_a values of 6.3 and 10.3, the bicarbonate ion is a stronger base than the carbonate ion

    C. Acetic acid ($pK_a = 4.7$) is stronger than lactic acid ($pK_a = 3.9$)

    D. Lactic acid ($pK_a = 3.9$) is weaker than all forms of phosphoric acid ($pK_a = 2.1, 6.9$ and 12.4)

    E. None of the above

188. What is the molarity of a sulfuric acid solution if 30.00 mL of H_2SO_4 is required to neutralize 0.840 g of sodium hydrogen carbonate (84.01 g/mol)?

$$H_2SO_4\,(aq) + 2\,NaHCO_3\,(aq) \rightarrow Na_2SO_4\,(aq) + 2\,H_2O\,(l) + 2\,CO_2\,(g)$$

    A. 0.333 M   **B.** 0.500 M   **C.** 0.167 M   **D.** 0.300 M   **E.** 0.667 M

189. Acetic acid is a weak acid in water because it is:

    A. only slightly dissociated into ions   **C.** only slightly soluble

    B. unable to hold onto its hydrogen ion   **D.** dilute

    E. completely dissociated into hydronium ions and acetate ions

190. Why is sulfuric acid a much stronger acid than carbonic acid?

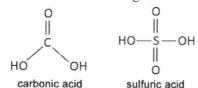

            carbonic acid   sulfuric acid

    A. The acid strength of two comparative molecules is directly proportional to the number of oxygens bonded to the central atom

    B. Since carbonic acid has resonance stabilization and sulfuric acid does not, sulfuric acid is less stable and more acidic

    C. The two double bonded oxygens in H_2SO_4 tend to destabilize the single bonded oxygen once the hydrogen ions form, thus making sulfuric acid more acidic

    D. In sulfuric acid, the negative charges can move between 4 oxygen atoms, compared to 3 oxygen atoms in carbonic acid

    E. In carbonic acid, the positive charge is able to move between two additional oxygens, rather than only one for sulfuric acid

191. What is the pH of a solution that has a hydronium ion concentration of 3.98×10^{-9} M?

   A. 8.000   **B.** 3.900   **C.** 5.750   **D.** 2,460   **E.** 8.400

192. What is the term for the product of an acid-base reaction in addition to water?

   A. nonelectrolyte   **B.** salt   **C.** buffer   **D.** electrolyte   **E.** none of the above

193. Which is the correct combination of Brønsted-Lowry bases in the following equilibrium?

$$H_2PO_4^- + H_2O \leftrightarrow H_3PO_4 + OH^-$$

   A. $H_2O + OH^-$ **C.** $H_2PO_4^- + H_3PO_4$

   B. $H_2O + H_3PO_4$ **D.** $H_2PO_4^- + OH^-$ **E.** $H_2PO_4^- + H_2O$

194. A buffer solution with pH = 4.0 and an acid with a pK_a of 3.0 should be prepared with a salt to acid ratio of:

   A. 1:1   **B.** 1:1000   **C.** 1:100   **D.** 1:5   **E.** 10:1

195. When a base is added to a buffered solution, the buffer:

   A. accepts H^+ **C.** releases H_2O

   B. releases protons   **D.** releases OH^- **E.** accepts H_3O^+

196. What is the term for a substance capable of either donating or accepting a proton in an acid-base reaction?

   A. semiprotic   **C.** aprotic

   B. nonprotic   **D.** amphiprotic   **E.** none of the above

197. For the reaction $C_5H_5N + H_2CO_3 \leftrightarrow C_5H_6N^+ + HCO_3^-$ the conjugate acid of C_5H_5N is:

   A. $C_5H_6N^+$ **B.** HCO_3^- **C.** C_5H_5N **D.** H_2CO_3 **E.** H_3O^+

198. Which species has a K_a of 6.6×10^{-10} if NH_3 has a K_b of 1.8×10^{-5}?

   A. H^+ **B.** NH_2^- **C.** NH_4^+ **D.** H_2O **E.** NH_3

199. The two members of the pair do NOT react for which of the following pairs of substances?

   A. Na_3PO_4 and HCl   **C.** HF and LiOH

   B. KCl and NaI   **D.** $PbCl_2$ and H_2SO_4 **E.** all react

200. If 20.0 mL of 0.500 M KOH is titrated with 0.250 M HNO_3, what volume of nitric acid is required to neutralize the base?

$$HNO_3\,(aq) + KOH\,(aq) \rightarrow KNO_3\,(aq) + H_2O\,(l)$$

   A. 10.0 mL   **B.** 20.0 mL   **C.** 25.0 mL   **D.** 40.0 mL   **E.** 50.0 mL

201. In a basic solution, the pH is […] and $[H_3O^+]$ is […].

   A. < 7 and $> 1 \times 10^{-7}$ M   **C.** $= 7$ and 1×10^{-7} M

   B. < 7 and $< 1 \times 10^{-7}$ M   **D.** < 7 and 1×10^{-7} M   **E.** > 7 and $< 1 \times 10^{-7}$ M

Copyright © 2015 Sterling Test Prep

202. What is the ratio of base/acid in the blood stream at pH 7.4 for formic acid (pK$_a$ = 3.9), which is the active agent in an ant bite?

    A. 3.5   **B.** 0.54   **C.** 3.16×10^{-4} **D.** 3.16×10^{3} **E.** 1.90

203. Which of the following is the ionization constant expression for water?

    A. $K_w = [H_2O] / [H^+] [OH^-]$ **C.** $K_w = [H^+] [OH^-]$

    B. $K_w = [H^+] [OH^-] / [H_2O]$ **D.** $K_w = [H_2O] [H_2O]$

                                                E. None of the above

204. Which reaction produces $NiCr_2O_7$ as a product?

    A. nickel (II) hydroxide and dichromic acid

    B. nickel (II) hydroxide and chromic acid

    C. nickelic acid and chromium (II) hydroxide

    D. nickel (II) hydroxide and chromate acid

    E. nickel (II) hydroxide and bichromic acid

205. Which of the following compounds cannot act as an acid?

    A. NH_3 **B.** H_2SO_4 **C.** HSO_4^{1-} **D.** SO_4^{2-} **E.** CH_3CO_2H

206. The pH of a solution for which $[H_3O^+] = 8.3 \times 10^{-9}$ is:

    A. 5.03   **B.** 11.54   **C.** 9.51   **D.** 5.88   **E.** 8.08

207. If a drain cleaner solution is a strong electrolyte, the drain cleaner is:

    A. slightly ionized   **C.** slightly reactive

    B. highly ionized   **D.** highly reactive

                                        E. None of the above

208. Calculate the $[H_3O^+]$ and the pH of a 0.021 M HNO_3 solution.

    A. 0.021 M and 1.68   **C.** 4.8×10^{-13} M and 12.32

    B. 0.021 M and −1.68   **D.** 4.8×10^{-13} M and −12.32

                                          E. 0.042 and 3.36

209. A sample of $Mg(OH)_2$ salt is dissolved in water and reaches equilibrium with its dissociated ions. Addition of the strong base NaOH increases the concentration of:

    A. H_2O^+ **C.** undissociated sodium hydroxide

    B. Mg^{2+} **D.** undissociated magnesium hydroxide

                                        E. H_2O^+ and undissociated sodium hydroxide

210. What is the molarity of a nitric acid solution, if 25.00 mL of HNO_3 is required to neutralize 0.500 g of calcium carbonate (100.09 g/mol)?

$$2\ HNO_3\,(aq) + CaCO_3\,(s) \rightarrow Ca(NO_3)_2\,(aq) + H_2O\,(l) + CO_2\,(g)$$

    A. 0.200 M   **B.** 0.250 M   **C.** 0.400 M   **D.** 0.550 M   **E.** 0.700 M

Copyright © 2015 Sterling Test Prep

211. Which of the following compounds is a salt?

A. NaOH **B.** H_2SO_4 **C.** $C_6H_{12}O_6$ **D.** HBr **E.** KNO_3

212. Boiler scale forms on the walls of hot water pipes from ground water due to:

A. transformation of $H_2PO_4^-$ ions to PO_4^{3-} ions, which precipitate with the "hardness ions", Ca^{2+}, Mg^{2+}, Fe^{2+}/Fe^{3+}

B. transformation of HSO_4^- ions to SO_4^{2-} ions, which precipitate with the "hardness ions", Ca^{2+}, Mg^{2+}, Fe^{2+}/Fe^{3+}

C. transformation of HSO_3^- ions to SO_3^{2-} ions, which precipitate with the "hardness ions", Ca^{2+}, Mg^{2+}, Fe^{2+}/Fe^{3+}

D. transformation of HCO_3^- ions to CO_3^{2-} ions, which precipitate with the "hardness ions", Ca^{2+}, Mg^{2+}, Fe^{2+}/Fe^{3+}

E. The reaction of the CO_3^{2-} ions present in all ground water with the "hardness ions", Ca^{2+}, Mg^{2+}, Fe^{2+}/Fe^{3+}

213. Which of the following statements about weak acids is correct?

A. Weak acids can only be prepared as dilute solutions

B. Weak acids have a strong affinity for acidic hydrogens

C. Weak acids always contain carbon atoms

D. The percentage dissociation for weak acids is usually in the range of 40-60%

E. Weak acid have a weak affinity for acidic hydrogens

214. Which of the following is an example of an Arrhenius acid?

A. H_2O *(l)* **C.** $Ba(OH)_2$ *(aq)*

B. RbOH *(aq)* **D.** $Al(OH)_3$ *(s)* **E.** None of the above

215. By the Arrhenius definition of acids and bases, a base:

A. bonds to H^+ **C.** decreases pH

B. decreases OH^- **D.** decreases H_2O **E.** increases H_3O^+

216. Which of the following is the strongest weak acid?

A. HF $pK_a = 3.17$ **C.** $H_2PO_4^-$ $pK_a = 7.18$

B. HCO_3^- $pK_a = 10.32$ **D.** NH_4^+ $pK_a = 9.20$

E. $HC_2H_3O_2$ $pK_a = 4.76$

217. If a light bulb in a conductivity apparatus glows dimly when testing a solution, which of the following must be true about the solution?

A. It is slightly reactive **C.** It is slightly ionized

B. It is highly reactive **D.** It is highly ionized **E.** None of the above

218. A Brønsted-Lowry base is a substance which:

A. donates protons to other substances **D.** produces hydroxide ions in aqueous solution

B. accepts protons from other substances **E.** accepts hydronium ions from other substances

C. produces hydrogen ions in aqueous solution

 Copyright © 2015 Sterling Test Prep

219. Which is a basic anhydride?

    A. BaO **B.** O_2 **C.** CO_2 **D.** SO_2 **E.** N_2O_5

220. A solution with a pH of 2.1 is:

    A. weakly basic   **C.** strongly acidic

    B. strongly basic   **D.** weakly acidic   **E.** a strong buffer

221. Which of the following indicators is green at pH 7?

    A. phenolphthalein   **C.** methyl red

    B. bromthymol blue   **D.** all of the above   **E.** none of the above

222. A Brønsted-Lowry base is defined as a substance that:

    A. increases $[H^+]$ when placed in water   **C.** acts as proton donor in any system

    B. decreases $[H^+]$ when placed in water   **D.** acts as proton acceptor in any system

                                       E. acts as a buffer

223. The K_a of formic acid ($HCOOH$) is 1.8×10^{-4}. What is the pK_b of the formate ion?

    A. $14 - \log(1.8 \times 10^{-4})$ **C.** $-14 - \log(1.8 \times 10^{-4})$

    B. $14 + \log(1.8 \times 10^{-10})$ **D.** $-14 + \log(1.8 \times 10^{-4})$ **E.** $14 + \log(1.8 \times 10^{-4})$

224. Which of the following indicators is yellow in a basic solution and red in an acidic solution?

    A. phenolphthalein   **C.** methyl red

    B. bromthymol blue   **D.** all of the above   **E.** none of the above

225. A typical amino acid has a carboxylic acid and an amine with pK_a values of 2.3 and 9.6, respectively. In a solution of pH 4.5, which describes its protonation and charge state?

    A. Carboxylic acid is deprotonated and negative; amine is protonated and neutral

    B. Carboxylic acid is deprotonated and negative; amine is protonated and positive

    C. Carboxylic acid is protonated and neutral; amine is deprotonated and negative

    D. Carboxylic acid is protonated and neutral; amine is protonated and neutral

    E. Carboxylic acid is deprotonated and negative; amine is deprotonated and neutral

226. Which substance will ionize when dissolved in water to form an aqueous solution?

    A. $C_6H_{12}O_6\,(s)$ **B.** $(NH_4)_2SO_4\,(s)$ **C.** $NaClO_4\,(s)$ **D.** $HNO_3\,(aq)$ **E.** $Ba(OH)_2\,(s)$

227. Identify a conjugate acid-base pair, whereby the acid is listed first in the pair.

$$H_2PO_4^- + S^{2-} \rightarrow HS^- + HPO_4^{2-}$$

Acid	Conjugate Base
A. $H_2PO_4^-$	HPO_4^{2-}
B. HS^-	$H_2PO_4^-$
C. S^{2-}	HS^-
D. $H_2PO_4^-$	S^{2-}
E. HPO_4^{2-}	$H_2PO_4^-$

228. What is the $[H^+]$ in stomach acid that registers a pH of 2 on a strip of pH paper?

    A. 0.2 M   **B.** 0.1 M   **C.** 0.02 M   **D.** 0.01 M   **E.** 2 M

229. Addition of sodium acetate to a solution of acetic acid causes the pH to:

    A. decrease due to the common ion effect

    B. increase due to the common ion effect

    C. remain constant because sodium acetate is a buffer

    D. remain constant because sodium acetate is neither acidic nor basic

    E. remain constant due to the common ion effect

230. Which of the following substances, when added to a solution of nitrous acid (HNO_2), could be used to prepare a buffer solution?

    A. H_2O **B.** $HC_2H_3O_2$ **C.** NaCl   **D.** HCl   **E.** $NaNO_2$

231. If 25.0 mL of 0.100 M $Ca(OH)_2$ is titrated with 0.200 M HNO_3, what volume of nitric acid is required to neutralize the base?

$$2HNO_3\,(aq) + Ca(OH)_2\,(aq) \rightarrow 2Ca(NO_3)_2\,(aq) + 2H_2O\,(l)$$

    A. 25.0 mL   **B.** 40.0 mL   **C.** 12.5 mL   **D.** 20.0 mL   **E.** 50.0 mL

232. What is the pK_a of an unknown acid if, in a solution at pH 7.0, 24% of the acid is in its deprotonated form?

    A. 6.0   **B.** 6.5   **C.** 7.5   **D.** 8.0   **E.** 10.0

233. Which statement is true to distinguish between dissociation and ionization?

    A. Some acids are weak electrolytes and ionize completely when dissolved in H_2O

    B. Some acids are strong electrolytes and dissociate completely when dissolved in H_2O

    C. Some acids are strong electrolytes and ionize completely when dissolved in H_2O

    D. All acids are strong electrolytes and dissociate completely when dissolved in H_2O

    E. Some acids are weak electrolytes and dissociate partially when dissolved in H_2O

234. Which species is formed in the second step of the dissociation of H_3PO_4?

    A. PO_4^{3-} **B.** $H_2PO_4^{2-}$ **C.** $H_2PO_4^-$ **D.** HPO_4^{2-} **E.** H_3PO_3

235. What are the products from the complete neutralization of phosphoric acid with aqueous lithium hydroxide?

    A. $LiHPO_4\,(aq)$ and $H_2O\,(l)$ **C.** $Li_2HPO4\,(aq)$ and $H_2O\,(l)$

    B. $Li_3PO_4\,(aq)$ and $H_2O\,(l)$ **D.** $LiH_2PO_4\,(aq)$ and $H_2O\,(l)$

                                                            E. $Li_2PO_4\,(aq)$ and $H_2O\,(l)$

Copyright © 2015 Sterling Test Prep

236. Which acid has the lowest boiling point elevation?

Monoprotic Acids	Ka
Acid I	1×10^{-8}
Acid II	1×10^{-9}
Acid III	3.5×10^{-10}
Acid IV	2.5×10^{-8}

A. I   **B.** II   **C.** III   **D.** IV   **E.** Not enough data to determine

237. Based upon the reduction potential: $Zn^{2+} + 2e^- \rightarrow Zn\ (s)$ $E° = -0.76$ V. Does a reaction take place when Zinc (s) is added to aqueous HCl, under standard conditions?
  A. Yes, because the reduction potential for H^+ is negative
  B. Yes, because the reduction potential for H^+ is zero
  C. No, because the oxidation potential for Cl^- is positive
  D. No, because the reduction potential for Cl^- is negative
  E. Yes, because the reduction potential for H^+ is positive

238. If a battery acid solution is a strong electrolyte, which of the following must be true of the battery acid?
  A. It is highly ionized
  B. It is slightly ionized
  C. It is highly reactive
  D. It is slightly reactive
  E. It is slightly ionized and weakly reactive

239. Which acid or base is least likely to be present in a biochemical reaction?
  A. ammonia
  B. citric acid
  C. nitric acid
  D. acetic acid
  E. phosphoric acid

240. Which is the acid anhydride for $HClO_4$?
  A. ClO **B.** ClO_2 **C.** ClO_3 **D.** ClO_4 **E.** Cl_2O_7

241. Which of the following does NOT represent a conjugate acid/base pair?
  A. $HCl\ /\ Cl^-$
  B. $HC_2H_3O_2\ /\ {}^-OH$
  C. $H_3O^+\ /\ H_2O$
  D. $HCN\ /\ CN^-$
  E. All represent a conjugate acid/base pair

242. Which of the following reactions represents the ionization of H_2O?
  A. $H_2O + H_2O \rightarrow 2H_2 + O_2$
  B. $H_2O + H_2O \rightarrow H_3O^+ + OH^-$
  C. $H_2O + H_3O^+ \rightarrow H_3O^+ + H_2O$
  D. $H_3O^+ + OH^- \rightarrow H_2O + H_2O$
  E. None of the above

243. What pH range is generally considered to be within the physiological pH range?
  A. 1.5–3.5   **B.** 4.5–7.3   **C.** 5.5–8.5   **D.** 6.5–8.0   **E.** 7.5–10.0

244. An acid-base neutralization is the reaction of:

A. H^+ (*aq*) with O_2 (*g*) to form H_2O (*l*)   C. H_2 (*g*) with O_2 (*g*) to form H_2O (*l*)

B. Na^+ (*aq*) with OH^- (*aq*) to form NaOH (*aq*)   D. H_2 (*aq*) with OH^- (*aq*) to form H_2O (*l*)

                                                            E. H^+ (*aq*) with OH^- (*aq*) to form H_2O (*l*)

245. In which of the following pairs of substances are both species salts?

A. NH_4F and KCl   C. LiOH and K_2CO_3

B. $CaCl_2$ and HCN   D. NaOH and HNO_3 E. HCN and K_2CO_3

246. If a buffer is made with the pH below the pK_a of the weak acid, the [base] / [acid] ratio is:

A. equal to 0   C. greater than 1

B. equal to 1   D. less than 1   E. cannot be determined

247. Which is NOT a strong acid?

A. $HClO_3$ B. HF   C. HBr   D. HCl   E. HI

248. A Brønsted-Lowry acid is defined as a substance that:

A. acts as a proton donor in any system   C. increases $[H^+]$ when placed in water

B. acts as a proton acceptor in any system   D. decreases $[H^+]$ when placed in water

                                                           E. increases $[H^+]$ in any system

249. Which of the following is the conjugate acid of hydrogen phosphate, HPO_4^{2-}?

A. $H_2PO_4^{2-}$ B. H_3PO_4 C. $H_2PO_4^-$ D. $H_2PO_3^-$ E. none of the above

250. Which statement is NOT correct?

A. Acidic salts are formed by partial neutralization of a diprotic acid by a diprotic base

B. Acidic salts are formed by partial neutralization of a triprotic acid by a diprotic base

C. Acidic salts are formed by partial neutralization of a monoprotic acid by a monoprotic base

D. Acidic salts are formed by partial neutralization of a diprotic acid by a monoprotic base

E. Acidic salts are formed by partial neutralization of a polyprotic acid by a monoprotic base

251. The pH of a solution for which $[^-OH] = 1.0 \times 10^{-9}$ is:

A. 8.00   B. 1.00×10^{-5} C. 1.50   D. 5.00   E. 9

252. Lysine contains two amine groups (pK_a 9.0 and 10.0) and a carboxylic acid group (pK_a 2.2). In a solution of pH 9.5, which describes the protonation and charge state of lysine?

A. Carboxylic acid is deprotonated and negative; amine (pK_a 9.0) is deprotonated and neutral whereby the amine (pK_a 10.0) is protonated and positive

B. Carboxylic acid is deprotonated and negative; amine (pK_a 9.0) is protonated and positive whereby the amine (pK_a 10.0) is deprotonated and neutral

C. Carboxylic acid is deprotonated and neutral; both amines are protonated and positive

D. Carboxylic acid is deprotonated and negative; both amines are deprotonated and neutral

E. Carboxylic acid is deprotonated and neutral; amine (pK_a 9.0) is deprotonated and neutral whereby the amine (pK_a 10.0) is protonated and positive

Copyright © 2015 Sterling Test Prep

253. The acidic anhydride of phosphoric acid (H_3PO_4) is:

    A. P_2O **B.** P_2O_3 **C.** PO_3 **D.** PO_2 **E.** P_4O_{10}

254. Which acid has the strongest conjugate base?

Monoprotic Acids	Ka
Acid I	1×10^{-8}
Acid II	1×10^{-9}
Acid III	3.5×10^{-10}
Acid IV	2.5×10^{-8}

    A. I   **B.** II   **C.** III   **D.** IV   **E.** Not enough data to determine

Copyright © 2015 Sterling Test Prep

Copyright © 2015 Sterling Test Prep

Chapter 6.   STOICHIOMETRY

1. Which substance listed below is the strongest oxidizing agent given the following spontaneous redox reaction?

$$Mg\ (s) + Sn^{2+}\ (aq) \rightarrow Mg^{2+}\ (aq) + Sn\ (s)$$

   A. Mg^{2+} **B.** Sn   **C.** Mg   **D.** Sn^{2+} **E.** None of the above

2. Which of the following is the correct conversion factor to convert mL to L?

   A. 1 L = 1000 mL   **C.** 1000 L = 1 mL

   B. 1 L = 100 mL   **D.** 10 L = 1 mL   **E.** 10 L = 1000 mL

3. How many grams are in 0.4 mole of $CaCO_3$?

   A. 25 g   **B.** 33 g   **C.** 40 g   **D.** 58 g   **E.** 67 g

4. How many moles of C are in a 3.50 g sample of C if the atomic mass of C is 12.011 amu?

   A. 3.61 moles   **C.** 1.00 moles

   B. 0.367 moles   **D.** 3.34 moles   **E.** 0.291 moles

5. Which reaction is NOT correctly classified?

   A. $PbO\ (s) + C\ (s) \rightarrow Pb\ (s) + CO\ (g)$ (double-replacement)

   B. $CaO\ (s) + H_2O\ (l) \rightarrow Ca(OH)_2\ (aq)$ (synthesis)

   C. $Pb(NO_3)_2\ (aq) + 2LiCl\ (aq) \rightarrow 2LiNO_3\ (aq) + PbCl_2\ (s)$ (double-replacement)

   D. $Mg\ (s) + 2HCl\ (aq) \rightarrow MgCl_2\ (aq) + H_2\ (g)$ (single-replacement)

   E. All are classified correctly

6. What is the molecular formula for lactic acid if the percent composition is 40.00% C, 6.71% H and 53.29% O, with an approximate molar mass of 90 g/mol?

   A. CHO_2 **B.** $C_3H_6O_3$ **C.** CHO   **D.** CH_2O **E.** C_6HO_8

7. Consider the balanced equation shown and identify the statement that is NOT true.

$$Na_2SO_4\ (aq) + BaCl_2\ (aq) \rightarrow 2NaCl\ (aq) + BaSO_4\ (s)$$

   A. The products are barium sulfate and sodium chloride

   B. Barium chloride is dissolved in water

   C. Barium sulfate is produced in solid form

   D. $2NaCl\ (aq)$ could also be written as $Na_2Cl_2\ (aq)$

   E. The coefficient of sodium sulfate is one

8. Which must be true concerning a solution at equilibrium when chemicals are mixed in a redox reaction and allowed to come to equilibrium?

   A. $\Delta G° = \Delta G$ **B.** $E = 0$ **C.** $\Delta G° = 0$ **D.** $K = 1$ **E.** $K < 1$

9. Balance the following chemical equation: ___N_2 + ___H_2 → ___NH_3

   A. 2, 6, 4   **B.** 3, 2, 1   **C.** 1, 2, 3   **D.** 1/2, 3/2, 1   **E.** 1, 3, 2

10. Which reaction is NOT correctly classified?

   A. $BaCl_2 + H_2SO_4 → BaSO_4 + 2HCl$ (single-replacement)
   B. $F_2 + 2NaCl → Cl_2 + 2NaF$ (single-replacement)
   C. $Fe + CuSO_4 → Cu + FeSO_4$ (single-replacement)
   D. $2NO_2 + H_2O_2 → 2HNO_3$ (synthesis)
   E. All are classified correctly

11. Which of the following nonmetals in the free state has an oxidation number of zero?

   A. S_8 **B.** P_4 **C.** F_2 **D.** Ar   **E.** All of the above

12. The number of moles of H_2O in a flask of water that contains 4.0×10^{21} molecules is:

   A. 2.4×10^{23} **C.** 6.6×10^{-3}
   B. 6.6×10^{-23} **D.** 2.4×10^{42} **E.** 2.4×10^{23}

13. Avogadro's number is:

   A. number of atoms in 1 g atomic weight of any element   **C.** approximately 6.02×10^{23}
   B. number of molecules in a compound   **D.** choices A and C
                  E. all of the above

14. What is the oxidation number of Cl in $HClO_4$?

   A. −1   **B.** +5   **C.** +4   **D.** +8   **E.** +7

15. Which conversion factor is NOT consistent with the equation below?

       $4NH_3 + 5O_2 → 4NO + 6H_2O$?

   A. 5 moles O_2 / 6 moles H_2O **C.** 4 moles NH_3 / 5 moles H_2O
   B. 4 moles NO / 4 moles NH_3 **D.** 4 moles NO / 5 moles O_2
                  E. All are classified correctly

16. What is the term for a list of the mass percent of each element in a compound?

   A. mass composition   **C.** compound composition
   B. percent composition   **D.** elemental composition
                  E. none of the above

17. In the following reaction which species is being oxidized and which is being reduced?

       $2Cr\ (s) + 3Cl_2\ (g) → 2CrCl_3\ (s)$

   A. Cr is oxidized, while Cl_2 is reduced   **C.** Cr is oxidized, while $CrCl_3$ is reduced
   B. Cl_2 is oxidized, while Cr is reduced   **D.** $CrCl_3$ is oxidized, while Cr is reduced
                  E. Cr is reduced, while $CrCl_3$ is reduced

 Copyright © 2015 Sterling Test Prep

18. If two different isotopes of an element are isolated as neutral atoms, the atoms must have the same number of:

      I. protons   II. neutrons   III. electrons

    A. I only   **B.** II only   **C.** I and III only   **D.** I, II and III   **E.** III only

19. What coefficient is needed for the O_2 molecule to balance the following equation?

$$2C_4H_{10}\,(g) + \underline{\quad}O_2\,(g) \rightarrow 8CO_2\,(g) + 10H_2O\,(l)$$

    A. 5   **B.** 1   **C.** 8   **D.** 15   **E.** 13

20. Choose the spectator ions: $Pb(NO_3)_2\,(aq) + H_2SO_4\,(aq) \rightarrow PbSO_4\,(s) + 2HNO_3\,(aq)$

    A. NO_3^- and H^+ **C.** Pb^{2+} and H^+

    B. H^+ and SO_4^{2-} **D.** Pb^{2+} and NO_3^- **E.** Pb^{2+} and SO_4^{2-}

21. What is the oxidation number of Br in $NaBrO_3$?

    A. −1   **B.** +1   **C.** +3   **D.** +5   **E.** None of the above

22. How many significant figures are there in the following number: 46,000 pounds?

    A. 1   **B.** 2   **C.** 3   **D.** 4   **E.** 5

23. Approximately what percent of $AgNO_3$ is oxygen by weight?

    A. 22%   **B.** 24%   **C.** 28%   **D.** 32%   **E.** 34%

24. How many moles of Mg are in a 3.50 g sample of Mg if the atomic mass of Mg is 24.305 amu?

    A. 0.0175 moles   **C.** 0.224 moles

    B. 0.144 moles   **D.** 0.232 moles   **E.** 1.23×10^{23} moles

25. Which reaction below is a *decomposition* reaction?

    A. $2Cr\,(s) + 3Cl_2\,(g) \rightarrow 2CrCl_3\,(s)$ **C.** $C_7H_8O_2\,(l) + 8O_2\,(g) \rightarrow 7CO_2\,(g) + 4H_2O\,(l)$

    B. $6Li\,(s) + N_2\,(g) \rightarrow 2Li_3N\,(s)$ **D.** $2KClO_3\,(s) \rightarrow 2KCl\,(s) + 3O_2\,(g)$

                                              E. None of the above

26. After balancing the following redox reaction, what is the coefficient of O_2?

$$Al_2O_3\,(s) + Cl_2\,(g) \rightarrow AlCl_3\,(aq) + O_2\,(g)$$

    A. 1   **B.** 2   **C.** 3   **D.** 5   **E.** None of the above

27. In a chemical reaction:

    A. there is always the same number of products as there are reactants

    B. there are equal numbers of atoms on each side of the reaction arrow

    C. the number of atoms present in a reaction can vary when the conditions change during the reaction

    D. there are equal numbers of molecules on each side of the reaction arrow

    E. none of the above

Copyright © 2015 Sterling Test Prep

28. A gas sample had $P = 2.0$ atm, $V = 10.5$ L and $T = 298$ K. Based on these measurements, an investigator calculated the number of moles from $PV = nRT$. Which value has the correct number of significant figures? (Note: $R = 0.08206$ L atm K^{-1} mol^{-1})

    A. 0.12 mol   **B.** 0.123 mol   **C.** 0.1227 mol   **D.** 0.12270 mol   **E.** 0.1 mol

29. For the balanced equation, which term has the highest coefficient: $4H_2 + 2C \rightarrow 2CH_4$?

    A. CH_4 **B.** H_4 **C.** H_2 **D.** C   **E.** None of the above

30. Which reaction below is a *synthesis* reaction?

    A. $3CuSO_4 + Al \rightarrow Al_2(SO_4)_3 + 3Cu$ **C.** $2NaHCO_3 \rightarrow Na_2CO_3 + CO_2 + H_2O$

    B. $SO_3 + H_2O \rightarrow H_2SO_4$ **D.** $C_3H_8 + 5O_2 \rightarrow 3CO_2 + 4H_2O$

                                                   E. None of the above

31. What is the oxidation number of metallic iron in the elemental state?

    A. 0   **B.** +1   **C.** +2   **D.** +3   **E.** None of the above

32. The reactants for this chemical reaction are:

$$C_6H_{12}O_6 + 6H_2O + 6O_2 \rightarrow 6CO_2 + 12H_2O$$

    A. $C_6H_{12}O_6$, H_2O, O_2 and CO_2 **C.** $C_6H_{12}O_6$

    B. $C_6H_{12}O_6$ and H_2O **D.** $C_6H_{12}O_6$ and CO_2

                                                 E. $C_6H_{12}O_6$, H_2O and O_2

33. What is the empirical formula of the compound with the following percent composition?

    $C = 15.8\%$ $S = 42.1\%$ $N = 36.8\%$ $H = 5.3\%$

    A. N_2H_4CS **C.** $N_2H_2C_3S$

    B. NH_2C_3S **D.** N_3H_2CS **E.** $NH_2C_3S_4$

34. The formula mass of $Co(NH_3)_6(ClO_4)_3$ is:

    A. 317.76 amu   **C.** 403.68 amu

    B. 384.63 amu   **D.** 459.47 amu   **E.** 751.74 amu

35. Which reaction is NOT correctly classified?

    A. $AgNO_3 + KCl \rightarrow KNO_3 + AgCl$ (double-replacement)

    B. $CH_4 + 2O_2 \rightarrow CO_2 + 2H_2O$ (single-replacement)

    C. $Zn + H_2SO_4 \rightarrow ZnSO_4 + H_2$ (single-replacement)

    D. $2KClO_3 \rightarrow 2KCl + 3O_2$ (decomposition)

    E. All are classified correctly

36. What is the oxidation number of liquid bromine in the elemental state?

    A. 0   **B.** −1   **C.** −2   **D.** −3   **E.** None of the above

                                        Copyright © 2015 Sterling Test Prep

37. Which of the following is the percent mass composition of acetic acid (CH_3COOH)?

    A. 48% carbon, 8% hydrogen and 44% oxygen

    B. 52% carbon, 12% hydrogen and 36% oxygen

    C. 32% carbon, 6% hydrogen and 62% oxygen

    D. 40% carbon, 7% hydrogen and 53% oxygen

    E. 34% carbon, 4% hydrogen and 62% oxygen

38. Which of the following has the greatest mass?

    A. 34 protons, 34 neutrons and 39 electrons   **C.** 35 protons, 34 neutrons and 37 electrons

    B. 34 protons, 35 neutrons and 37 electrons   **D.** 34 protons, 35 neutrons and 34 electrons

    E. 35 protons, 35 neutrons and 33 electrons

39. Which equation(s) is/are balanced?

    I. $Mg\ (s) + 2HCl\ (aq) \rightarrow MgCl_2\ (aq) + H_2\ (g)$

    II. $3Al\ (s) + 3Br_2\ (l) \rightarrow Al_2Br_3\ (s)$

    III. $2HgO\ (s) \rightarrow 2\ Hg\ (l) + O_2\ (g)$

    A. Only equation III is balanced   **C.** Equations I and III are balanced

    B. Equations II and III are balanced   **D.** All equations are balanced

    E. None of the equations are balanced

40. Which of the following is a *double-replacement* reaction?

    A. $2HI \rightarrow H_2 + I_2$ **C.** $HBr + KOH \rightarrow H_2O + KBr$

    B. $SO_2 + H_2O \rightarrow H_2SO_3$ **D.** $CuO + H_2 \rightarrow Cu + H_2O$ **E.** None of the above

41. Which metal in the free state has an oxidation number of zero?

    A. Mg   **B.** Na   **C.** Al   **D.** Ag   **E.** All of the above

42. Which contains the fewest moles?

    A. 10g CH_4 **B.** 10g Si   **C.** 10g CO   **D.** 10g N_2 **E.** 10g AlH_3

43. How many moles are in 40.1g of $MgSO_4$?

    A. 0.25 mole   **B.** 0.33 mole   **C.** 0.45 mole   **D.** 0.55 mole   **E.** 0.67 mole

44. What is the oxidation number of sulfur in the $S_2O_8^{2-}$ ion?

    A. −1   **B.** +7   **C.** +2   **D.** +6   **E.** +1

45. By definition, a strong electrolyte must:

    A. be highly soluble in water

    B. contain both metal and nonmetal atoms

    C. be an ionic compound

    D. dissociate almost completely into its ions in solution

    E. none of the above

46. What is the oxidation number of Cl in $LiClO_2$?

   A. −1   **B.** +1   **C.** +3   **D.** +5   **E.** None of the above

47. The oxidation number of sulfur in calcium sulfate, $CaSO_4$, is:

   A. +6   **B.** +4   **C.** +2   **D.** 0   **E.** −2

48. If A of element Y equals 13, then 26 grams of element Y represents approximately:

   A. 26 moles of atoms   **C.** 2 atoms
   B. ½ mole of atoms   **D.** ½ of an atom   **E.** 2 moles of atoms

49. For the balanced reaction $2Na + Cl_2 \rightarrow 2NaCl$, which of the following is a solid?

   A. Cl_2 **B.** Cl   **C.** Na   **D.** NaCl   **E.** None are solids

50. Which of the following would be a weak electrolyte in a solution?

   A. HBr (*aq*)   **B.** KCl   **C.** KOH   **D.** $HC_2H_3O_2$ **E.** HI

51. What substance is the oxidizing agent in the following redox reaction?

$$Co\ (s)\ + 2\ HCl\ (aq) \rightarrow CoCl_2\ (aq)\ + H_2\ (g)$$

   A. H_2 **B.** $CoCl_2$ **C.** HCl   **D.** Co   **E.** None of the above

52. Which of the following conversion factors is correct for converting from grams to kilograms?

   A. 1000 g = 1 kg   **C.** 1 g = 100 kg
   B. 100 g = 1 kg   **D.** 1 g = 1000 kg   **E.** 10 g = 1 kg

53. If the molecular weight of a compound is 219 g/mol, what is the molecular formula of the compound with the following percent composition?

   C = 49.3%   O = 43.8%   H = 6.9%

   A. $C_3O_2H_5$ **B.** $C_2O_2H_5$ **C.** $C_3O_2H_3$ **D.** $C_3O_2H_4$ **E.** None of the above

54. The formula mass of $C_{14}H_{28}(COOH)_2$ is:

   A. 237.30 amu   **C.** 271.24 amu
   B. 252.32 amu   **D.** 286.41 amu   **E.** 271.50 amu

55. The net ionic equation for the reaction $CaCO_3 + 2HNO_3 \rightarrow Ca(NO_3)_2 + CO_2 + H_2O$ is:

   A. $CO_3^{2-} + H^+ \rightarrow CO_2$ **C.** $Ca^{2+} + 2\ NO_3^- \rightarrow Ca(NO_3)_2$
   B. $CaCO_3 + 2\ H^+ \rightarrow Ca^{2+} + CO_2 + H_2O$ **D.** $CaCO_3 + 2\ NO_3^- \rightarrow Ca(NO_3)_2 + CO_3^{2-}$
                          E. none of the above

56. What substance is reduced in the following redox reaction?

$$F_2\ (g) + 2\ Br^-\ (aq) \rightarrow 2\ F^-\ (aq) + Br_2\ (l)$$

   A. F^- **B.** Br_2 **C.** F_2 **D.** Br^- **E.** None of the above

                                   Copyright © 2015 Sterling Test Prep

57. Which is the correct equation for the reaction of magnesium with hydrochloric acid to yield hydrogen and magnesium chloride?

A. $2Mg + 6HCl \rightarrow 3H_2 + 2MgCl_2$

B. $Mg + 2HCl \rightarrow 2H + MgCl_2$

C. $Mg + 3HCl \rightarrow 3H + MgCl_2$

D. $Mg + HCl \rightarrow H + MgCl$

E. $Mg + 2HCl \rightarrow H_2 + MgCl_2$

58. Which represents the balanced double displacement reaction between copper (II) chloride and iron (II) carbonate?

A. $Cu_2Cl + FeCO_3 \rightarrow Cu_2CO_3 + FeC$

B. $CuCl_2 + FeCO_3 \rightarrow CuCO_3 + FeCl_2$

C. $Cu_2Cl + Fe_2CO_3 \rightarrow Cu_2CO_3 + Fe_2Cl$

D. $CuCl_2 + Fe_2(CO_3)_2 \rightarrow Cu(CO_3)_2 + FeCl_2$

E. $CuCl_2 + 2FeCO_3 \rightarrow 2CuCO_3 + FeCl_2$

59. In a chemical equation, the coefficients:

A. appear before the chemical formulas

B. appear as subscripts

C. reactants always sum up to those of the products

D. two of the above statements are correct

E. none of the above statements is correct

60. In which of the following pairs of substances would both species in the pair be written in molecular form in a net ionic equation?

A. NH_4Cl and $NaCl$

B. CO_2 and H_2SO_4

C. $LiOH$ and H_2

D. HF and CO_2

E. None of the above

61. What substance is the reducing agent in the following redox reaction?

$$Co\ (s) + 2HCl\ (aq) \rightarrow CoCl_2\ (aq) + H_2\ (g)$$

A. $CoCl_2$ **B.** H_2 **C.** Co **D.** HCl **E.** None of the above

62. Which is a true statement about H_2O as it begins to freeze?

A. Hydrogen bonds break

B. Number of hydrogen bonds decreases

C. Covalent bond strength increases

D. Molecules move closer together

E. Number of hydrogen bonds increases

63. What is the sum of the coefficients in the balanced reaction (no fractional coefficients)?

$$RuS\ (s) + O_2 + H_2O \rightarrow Ru_2O_3\ (s) + H_2SO_4$$

A. 23   **B.** 18   **C.** 13   **D.** 25   **E.** 29

64. How many formula units of NaCl are in 146 grams of sodium chloride?

A. 3.82×10^{-22}

B. 2.34×10^{24}

C. 1.50×10^{24}

D. 7.45×10^{25}

E. 4.04×10^{23}

65. In which of the following pairs of substances would both species in the pair be written in ionic form in a net ionic equation?

   A. $AgCl$ and CO_2 **C.** CH_3COOH and HNO_3

   B. KBr and NH_3 **D.** Na_2CO_3 and $Ba(NO_3)_2$ **E.** None of the above

66. What is the oxidation number of I in KIO_4?

   A. −1   **B.** +1   **C.** +3   **D.** +5   **E.** None of the above

67. What is the percent mass composition of H_2SO_4?

   A. 48% oxygen, 50% sulfur and 2% hydrogen

   B. 65% oxygen, 33% sulfur and 2% hydrogen

   C. 75% oxygen, 24% sulfur and 1% hydrogen

   D. 85% oxygen, 14% sulfur and 1% hydrogen

   E. 48% oxygen, 40% sulfur and 12% hydrogen

68. How many atoms of Mg are in a solid 48 g sample of magnesium?

   A. $2 \times 6.02 \times 10^{23}$ **C.** 2,400,000

   B. 4,800   **D.** $4 \times 6.02 \times 10^{23}$ **E.** 6.02×10^{23}

69. For the balanced reaction $2Na + Cl_2 \rightarrow 2NaCl$, which of the following is a gas?

   A. Cl   **B.** NaCl   **C.** Na   **D.** Cl_2 **E.** None are a gas

70. Which of the following is NOT an electrolyte?

   A. NaBr   **B.** Ne   **C.** KOH   **D.** HCl   **E.** HI

71. What substance is oxidized in the following redox reaction?

$$F_2(g) + 2\,Br^-(aq) \rightarrow 2\,F^-(aq) + Br_2(l)$$

   A. Br_2 **B.** F^- **C.** Br^- **D.** F_2 **E.** None of the above

72. How many mL of solution are there in 0.0500 L?

   A. 0.0000500 mL   **B.** 500 mL   **C.** 0.50 mL   **D.** 50.0 mL   **E.** 5.0 mL

73. Assuming STP, if 49 g of H_2SO_4 are produced in the following reaction, what volume of O_2 must be used in the reaction?

$$RuS(s) + O_2 + H_2O \rightarrow Ru_2O_3(s) + H_2SO_4$$

   A. 20.6 liters   **C.** 28.3 liters

   B. 31.2 liters   **D.** 29.1 liters   **E.** 25.2 liters

74. What is the total number of moles of gas in a sample that contains 16.0 g of CH_4, 16.0 g of O_2, 16.0 g of SO_2 and 33.0 g of CO_2?

   A. 2.31 moles   **C.** 2.63 moles

   B. 2.50 moles   **D.** 3.10 moles   **E.** 4.20 moles

 Copyright © 2015 Sterling Test Prep

75. The net ionic equation for the reaction $Ca(OH)_2 + 2HCl \rightarrow 2H_2O + CaCl_2$ is:

   A. $OH^- + H^+ \rightarrow H_2O$ **C.** $Ca^{2+} + 2Cl^- \rightarrow CaCl_2$

   B. $2OH^- + 2HCl \rightarrow 2H_2O$ **D.** $Ca(OH)_2 + 2H^+ \rightarrow Ca^{2+} + H_2O$

                                        E. None of the above

76. What is the term that expresses the ratio of mass per unit volume for a gas?

   A. molar volume   **C.** gas ratio

   B. molar mass   **D.** gas density   **E.** none of the above

77. What is the oxidizing agent, and why is it, in the reaction shown:

$$Ni\ (s) + CuCl_2\ (aq) \rightarrow Cu\ (s) + NiCl_2\ (aq)$$

   A. $CuCl_2$ – causes reduction   **C.** Ni – causes reduction

   B. $CuCl_2$ – is reduced   **D.** Ni – is reduced   **E.** $NiCl_2$ – is reduced

78. What is the mass of N in a 11.2 liter container when the partial pressure of nitrogen gas is 0.5 atmospheres at 25°C?

   A. 4.5g   **B.** 24g   **C.** 11g   **D.** 22g   **E.** 7g

79. Balance the following equation: ___NO → ___N_2O + ___NO_2

   A. 4, 4, 8   **B.** 1, 2, 4   **C.** 3, 1, 1   **D.** 3, 0, 0   **E.** 6, 2, 1

80. What is the oxidation number of the pure element zinc and a compound like $ZnSO_4$, respectively?

   A. +1 and 0   **B.** 0 and +2   **C.** 0 and 0   **D.** 0 and +1   **E.** +1 and +2

81. The formula for the illegal drug cocaine is $C_{17}H_{21}NO_4$ (303.39 g/mol). What is the percentage of oxygen in cocaine?

   A. 21.09%   **B.** 6.35%   **C.** 4.57%   **D.** 4.74%   **E.** 62.83%

82. How long is 1cm?

   A. 0.01 mm   **B.** 1 mm   **C.** 0.1 mm   **D.** 100 mm   **E.** 10 mm

83. If the following reaction is run at STP with excess H_2O, and 22.4 liters of O_2 react with 67 g of RuS, how many grams of H_2SO_4 are produced?

   $RuS\ (s) + O_2 + H_2O \rightarrow Ru_2O_3\ (s) + H_2SO_4$

   A. 28 g   **B.** 32 g   **C.** 44 g   **D.** 54 g   **E.** 58 g

84. How many oxygen atoms are in 3.62 g of fructose ($C_6H_{12}O_6$)?

   A. 1.34×10^{-21} **C.** 3.28×10^{24}

   B. 7.26×10^{22} **D.** 1.16×10^{23} **E.** 6.85×10^{20}

Copyright © 2015 Sterling Test Prep

85. Which species is NOT written as its constituent ions when the equation is expanded into the ionic equation?

$$Mg(OH)_2 \ (s) + 2HCl \ (aq) \rightarrow MgCl_2 \ (aq) + 2H_2O \ (l)$$

A. $Mg(OH)_2$ only

B. H_2O and $Mg(OH)_2$

C. HCl

D. $MgCl_2$

E. HCl and $MgCl_2$

86. What substance is oxidized in the following redox reaction?

$$HgCl_2 \ (aq) + Sn^{2+} \ (aq) \rightarrow Sn^{4+} \ (aq) + Hg_2Cl_2 \ (s) + Cl^- \ (aq)$$

A. Hg_2Cl_2 **B.** Sn^{4+} **C.** Sn^{2+} **D.** $HgCl_2$ **E.** None of the above

87. Which of the following equations is NOT balanced?

A. $SO_2 + O_2 \rightarrow SO_3$

B. $2H_2 + O_2 \rightarrow 2H_2O$

C. $C_3H_8 + 5O_2 \rightarrow 3CO_2 + 4H_2O$

D. $2Na + 2H_2O \rightarrow 2NaOH + H_2$

E. $2Al + 6HCl \rightarrow 2AlCl_3 + 3H_2$

88. What is the total charge of all the electrons in grams of He, whereby the charge on one mole of electrons is given by Faraday's constant ($F = 96,500$ C/mol)?

A. 48,250 C

B. 96,500 C

C. 193,000 C

D. 386,000 C

E. Cannot be determined with the information provided

89. What is the approximate formula mass of sulfur dioxide (SO_2)?

A. 28 amu   **B.** 36 amu   **C.** 62 amu   **D.** 64 amu   **E.** 68 amu

90. In the following reaction, H_2SO_4 is the:

$$H_2SO_4 + HI \rightarrow I_2 + SO_2 + H_2O$$

A. reducing agent and is reduced

B. reducing agent and is oxidized

C. oxidizing agent and is reduced

D. oxidizing agent and is oxidized

E. oxidizing agent, but is neither oxidized nor reduced

91. How many formula units of lithium iodide (LiI) have a mass equal to 4.24 g?

A. 2.55×10^{24} formula units

B. 5.24×10^{23} formula units

C. 1.90×10^{25} formula units

D. 5.24×10^{26} formula units

E. 1.91×10^{22} formula units

92. During exercise, perspiration on a person's skin can form droplets because of the:

A. ability of H_2O to dissipate heat

B. high specific heat of H_2O

C. adhesive properties of H_2O

D. cohesive properties of H_2O

E. high NaCl content of perspiration

Copyright © 2015 Sterling Test Prep

93. If 7 moles of RuS are used in the following reaction, what is the maximum number of moles of Ru_2O_3 that can be produced?

$$RuS \ (s) + O_2 + H_2O \rightarrow Ru_2O_3 \ (s) + H_2SO_4$$

   A. 3.5 moles   **C.** 3.2 moles

   B. 2.8 moles   **D.** 2.24 moles   **E.** 3.8 moles

94. The mass of 5.20 moles of glucose ($C_6H_{12}O_6$) is:

   A. 1.54×10^{-21} g   **C.** 344 g

   B. 313 g   **D.** 937 g   **E.** 6.34×10^{20} g

95. In which sequence of sulfur-containing species are the species arranged in *decreasing* oxidation numbers for S?

   A. SO_4^{2-}, S^{2-}, $S_2O_3^{2-}$ **C.** SO_4^{2-}, $S_2O_3^{2-}$, S^{2-}

   B. $S_2O_3^{2-}$, SO_3^{2-}, S^{2-} **D.** SO_3^{2-}, SO_4^{2-}, S^{2-} **E.** S^{2-}, SO_4^{2-}, $S_2O_3^{2-}$

96. What law is illustrated if ethyl alcohol always contains 52% carbon, 13% hydrogen, and 35% oxygen by mass?

   A. law of constant composition   **C.** law of multiple proportions

   B. law of constant percentages   **D.** law of conservation of mass

                                        E. none of the above

97. If one mole of Ag is produced, how many grams of O_2 gas are produced: $2Ag_2O \rightarrow 4Ag + O_2$?

   A. 6g   **B.** 5g   **C.** 2g   **D.** 12g   **E.** 8g

98. Silicon exists as three isotopes: ^{28}Si, ^{29}Si, and ^{30}Si with atomic masses of 27.98 amu, 28.98 amu and 29.97 amu, respectively. Which isotope is the most abundant in nature?

   A. ^{28}Si   **C.** ^{30}Si

   B. ^{29}Si   **D.** ^{28}Si are ^{30}Si equally abundant   **E.** All are equally abundant

99. Is it possible to have a macroscopic sample of oxygen that has a mass of 14 atomic mass units?

   A. No, because oxygen is a gas at room temperature

   B. Yes, because it would have the same density as nitrogen

   C. No, because this is less than a macroscopic quantity

   D. Yes, but it would need to be made of isotopes of oxygen atoms

   E. No, because this is less than the mass of a single oxygen atom

100. The oxidation numbers for the elements in Na_2CrO_4 would be:

   A. +2 for Na, +5 for Cr and −6 for O   **C.** +1 for Na, +4 for Cr and −6 for O

   B. +2 for Na, +3 for Cr and −2 for O   **D.** +1 for Na, +6 for Cr and −2 for O

                                              E. +1 for Na, +5 for Cr and −2 for O

101. After balancing the following redox reaction, what is the coefficient of NaCl?

$$Cl_2 \ (g) + NaI \ (aq) \rightarrow I_2 \ (s) + NaCl \ (aq)$$

A. 1   **B.** 2   **C.** 3   **D.** 5   **E.** none of the above

102. What is the mass of 3.61 moles of Ca?

A. 43.0g   **B.** 145g   **C.** 0.080g   **D.** 144g   **E.** 124g

103. When aluminum metal reacts with ferric oxide (Fe_2O_3), a displacement reaction yields two products with one of the products being metallic iron. What is the sum of the coefficients of the products of the balanced reaction?

A. 4   **B.** 6   **C.** 2   **D.** 5   **E.** 3

104. How many molecules of CO_2 are in 154.0 grams of CO_2?

A. 3.499   **C.** 4.214×10^{24}

B. 2.107×10^{24} **D.** 9.274×10^{25} **E.** 4.081×10^{27}

105. The oxidation number +7 is for the element:

A. Mn in $KMnO_4$ **C.** C in MgC_2O_4

B. Br in $NaBrO_3$ **D.** S in H_2SO_4 **E.** K in $KMnO_4$

106. Which substance listed below is the strongest reducing agent, given the following *spontaneous* redox reaction?

$$FeCl_3 \ (aq) + NaI \ (aq) \rightarrow I_2 \ (s) + FeCl_2 \ (aq) + NaCl \ (aq)$$

A. $FeCl_2$ **B.** I_2 **C.** NaI   **D.** $FeCl_3$ **E.** NaCl

107. Which statement regarding balanced chemical equations is NOT true?

A. When no coefficient is written in front of a formula, the number "one" is assumed

B. Subscripts may be changed to make an equation simpler to balance

C. Coefficients are used in front of formulas to balance the equation

D. The number of each kind of atoms must be the same on each side

E. Reactants are written to the left of the arrow

108. In the following reaction, which is performed at 600 K, 9.0 moles of N_2 gas are mixed with 22 moles of H_2 gas. What is the percent yield of NH_3 if the reaction produces 12 moles of NH_3?

$$N_2 \ (g) + 3H_2 \ (g) \rightarrow 2NH_3 \ (g)$$

A. 66%   **B.** 30%   **C.** 20%   **D.** 100%   **E.** 82%

109. Seven grams of N_2 contain:

A. 0.5 moles of nitrogen atoms   **C.** 0.25 moles of nitrogen atoms

B. 1 mole of nitrogen atoms   **D.** 0.75 moles of nitrogen atoms

     E. Not enough information is given

 Copyright © 2015 Sterling Test Prep

110. Which substance is functioning as the oxidizing agent in this reaction?

$$14\,H^+ + Cr_2O_7^{2-} + 3\,Ni \rightarrow 3\,Ni^{2+} + 2\,Cr^{3+} + 7\,H_2O$$

A. H_2O **B.** $Cr_2O_7^{2-}$ **C.** H^+ **D.** Ni   **E.** Ni^{2+}

111. What is the term for the value corresponding to the number of atoms in 12.01g of carbon?

A. mass number   **C.** Avogadro's number

B. mole number   **D.** atomic number   **E.** none of the above

112. How many atoms of oxygen does this reaction yield?

$$C_6H_{12}O_6 + 6H_2O + 6O_2 \rightarrow 6CO_2 + 12H_2O$$

A. 3   **B.** 12   **C.** 14   **D.** 24   **E.** 36

113. From the following reaction, if 0.1 mole of Al is allowed to react with 0.2 mole of Fe_2O_3, how many moles of iron are produced: $2Al + Fe_2O_3 \rightarrow 2Fe + Al_2O_3$

A. 0.05 mole   **C.** 0.10 mole

B. 0.075 mole   **D.** 0.15 mole   **E.** 0.25 mole

114. What is the balanced reaction for the combustion of methane?

A. $CH_4 + OH^- \rightarrow CH_3OH$ **C.** $NH_3 + OH^- \rightarrow NH_4OH$

B. $CH_4 + \frac{1}{2}O_2 \rightarrow CO_2 + 2H_2O$ **D.** $CH_3OH + 2O_2 \rightarrow CO_2 + 2H_2O$

E. $CH_4 + 2O_2 \rightarrow CO_2 + 2H_2O$

115. Which of the following represents 1 mol of phosphine gas (PH_3)?

A. 22.4 L phosphine gas at STP   **C.** 6.02×10^{23} phosphine molecules

B. 34.00 g phosphine gas   **D.** All of the above

E. None of the above

116. Select the balanced chemical equation:

A. $2C_2H_5OH + 2Na_2Cr_2O_7 + 8H_2SO_4 \rightarrow 2HC_2H_3O_2 + 2Cr_2(SO_4)_3 + 4Na_2SO_4 + 11H_2O$

B. $2C_2H_5OH + Na_2Cr_2O_7 + 8H_2SO_4 \rightarrow 3HC_2H_3O_2 + 2Cr_2(SO_4)_3 + 2Na_2SO_4 + 11H_2O$

C. $C_2H_5OH + 2Na_2Cr_2O_7 + 8H_2SO_4 \rightarrow HC_2H_3O_2 + 2Cr_2(SO_4)_3 + 2Na_2SO_4 + 11H_2O$

D. $C_2H_5OH + Na_2Cr_2O_7 + 2H_2SO_4 \rightarrow HC_2H_3O_2 + Cr_2(SO_4)_3 + 2Na_2SO_4 + 11H_2O$

E. $3C_2H_5OH + 2Na_2Cr_2O_7 + 8H_2SO_4 \rightarrow 3HC_2H_3O_2 + 2Cr_2(SO_4)_3 + 2Na_2SO_4 + 11H_2O$

117. The oxidation number of Cr in $K_2Cr_2O_7$ is:

A. +6   **B.** +5   **C.** +4   **D.** +2   **E.** +1

118. What is the volume occupied by 10.0 g of NO gas, at STP?

A. 13.4 L   **B.** 67.2 L   **C.** 0.333 L   **D.** 7.46 L   **E.** 224 L

119. What is the empirical formula of a compound that, by mass, contains 64% silver, 8% nitrogen and 28% oxygen?

    A. Ag_3NO **B.** Ag_3NO_3 **C.** $AgNO_2$ **D.** Ag_3NO_2 **E.** $AgNO_3$

120. Which reaction represents the balanced reaction for the combustion of ethanol?

    A. $4C_2H_5OH + 13O_2 \rightarrow 8CO_2 + 10H_2$

    B. $C_2H_5OH + 3O_2 \rightarrow 2CO_2 + 3H_2O$

    C. $C_2H_5OH + 2O_2 \rightarrow 2CO_2 + 2H_2O$

    D. $C_2H_5OH + O_2 \rightarrow CO_2 + H_2O$

    E. $C_2H_5OH + \frac{1}{2} O_2 \rightarrow 2CO_2 + 3H_2O$

121. What is the formula mass of a molecule of CO_2?

    A. 44 amu   **C.** 56.5 amu

    B. 52 amu   **D.** 112 amu   **E.** None of the above

122. What is the oxidizing agent in the redox reaction: $Al + MnO_2 \rightarrow Al_2O_3 + Mn$?

    A. O in Al_2O_3 **B.** O in MnO_2 **C.** Mn in MnO_2 **D.** Al   **E.** Al in Al_2O_3

123. What substance is reduced in the following redox reaction?

$$HgCl_2 \ (aq) + Sn^{2+} \ (aq) \rightarrow Sn^{4+} \ (aq) + Hg_2Cl_2 \ (s) + Cl^- \ (aq)$$

    A. Sn^{4+} **B.** Hg_2Cl_2 **C.** $HgCl_2$ **D.** Sn^{2+} **E.** None of the above

124. Convert 152 miles into kilometers, using significant figures, given that 1 mile = 1.609 km:

    A. 245 km   **B.** 244.57 km   **C.** 94 km   **D.** 94.4 km   **E.** 244.6 km

125. In acidic conditions, what is the sum of the coefficients in the products of the balanced reaction?

$$MnO_4^- + C_3H_7OH \rightarrow Mn^{2+} + C_2H_5COOH$$

    A. 12   **B.** 16   **C.** 18   **D.** 20   **E.** 24

126. What is the percent by mass of chromium in K_2CrO_4?

    A. 26.776 %   **B.** 31.663 %   **C.** 41.348 %   **D.** 43.765 %   **E.** 52.446 %

127. The oxidation number of C in BaC_2O_4 is:

    A. −3   **B.** −2   **C.** +2   **D.** +3   **E.** +4

128. How many molecules of CH_4 gas have a mass equal to 3.20 g?

    A. 3.01×10^{23} molecules   **C.** 1.20×10^{24} molecules

    B. 1.93×10^{24} molecules   **D.** 3.01×10^{24} molecules   **E.** 1.20×10^{23} molecules

129. When a substance loses electrons it is [   ], while the substance itself is acting as [   ] agent.

    A. reduced… a reducing   **C.** reduced… an oxidizing

    B. oxidized… a reducing   **D.** oxidized… an oxidizing   **E.** dissolved… a neutralizing

 Copyright © 2015 Sterling Test Prep

130. At constant volume, as the temperature of a sample of gas is decreased, the gas deviates from ideal behavior. Compared to the pressure predicted by the ideal gas law, actual pressure would be:

A. higher, because of the volume of the gas molecules

B. higher, because of intermolecular attractions between gas molecules

C. lower, because of the volume of the gas molecules

D. lower, because of intermolecular attractions among gas molecules

E. higher, because of intramolecular attractions between gas molecules

131. How many O_2 molecules are needed to yield $10CO_2$ molecules according to balanced chemical equation: $2CO + O_2 \rightarrow 2CO_2$

A. 4   **B.** 10   **C.** 5   **D.** 1   **E.** 2

132. In which of the following compounds does Cl have an oxidation number of +7?

A. $NaClO_2$ **C.** $Ca(ClO_3)_2$

B. $Al(ClO_4)_3$ **D.** $LiClO_3$ **E.** None of the above

133. Which of the following is a guideline for balancing redox equations by the oxidation number method?

A. Verify that the total number of atoms and the total ionic charge are the same for reactants and products

B. In front of the substance reduced, place a coefficient that corresponds to the number of electrons lost by the substance oxidized

C. In front of the substance oxidized, place a coefficient that corresponds to the number of electrons gained by the substance reduced

D. Diagram the electrons lost by the substance oxidized and gained by the substance reduced

E. All of the above

134. What is the coefficient for O_2 when the following equation is balanced with the lowest whole number coefficients?

$$C_3H_7OH + O_2 \rightarrow CO_2 + H_2O$$

A. 4   **B.** 5   **C.** 9   **D.** 10   **E.** 12

135. From the following reaction, if 0.1 mole Al is allowed to react with 0.2 mole Fe_2O_3, what is the limiting reactant in this reaction: $2Al + Fe_2O_3 \rightarrow 2Fe + Al_2O_3$

A. Fe_2O_3 **B.** Al_2O_3 **C.** Fe_2O_3 and Al_2O_3 **D.** Fe   **E.** Al

136. How many moles of Al are in a 4.56 g sample of Al if the atomic mass of Al is 26.982 amu?

A. 0.169 moles   **C.** 1.43 moles

B. 0.230 moles   **D.** 4.34 moles   **E.** 5.87 moles

137. Identify the missing terms in the following definition: "An oxidation number is the [   ] that an atom [   ] when the electrons in each bond are assigned to the [   ] electronegative of the two atoms involved in the bond.

    A. charge… definitely has… more

    B. charge… definitely has… less

    C. number of electrons… definitely has… more

    D. number of electrons… appears to have… less

    E. charge… appears to have… more

138. Which substance listed below is the weakest oxidizing agent given the following spontaneous redox reaction?

$$Mg\ (s) + Sn^{2+}\ (aq) \rightarrow Mg^{2+}\ (aq) + Sn\ (s)$$

    A. Mg^{2+} **B.** Sn   **C.** Mg   **D.** Sn^{2+} **E.** None of the above

139. The scientific principle that is the basis for balancing chemical equations is:

    A. Law of Conservation of Mass and Energy

    B. Law of Definite Proportions

    C. Law of Conservation of Energy

    D. Law of Conservation of Mass

    E. Avogadro's Law

140. How many moles of O_2 gas are required for combustion with one mole of $C_6H_{12}O_6$ in the unbalanced reaction: $C_6H_{12}O_6\ (s) + O_2\ (g) \rightarrow CO_2\ (g) + H_2O\ (g)$

    A. 1   **B.** 2.5   **C.** 6   **D.** 10   **E.** 12

141. What are the formula masses of water (H_2O), propene (C_3H_6) and 2-propanol (C_3H_8O)?

    A. water: 18 amu; propene: 44 amu; 2-propanol: 64 amu

    B. water: 18 amu; propene: 42 amu; 2-propanol: 60 amu

    C. water: 18 amu; propene: 42 amu; 2-propanol: 62 amu

    D. water: 18 amu; propene: 40 amu; 2-propanol: 58 amu

    E. water: 18 amu; propene: 42 amu; 2-propanol: 58 amu

142. Which of the following net ionic equations represents a disproportionation reaction?

    A. $ClO^- + Cl^- + 2H^+ \rightarrow Cl_2 + H_2O$

    B. $CH_4 + 2O_2 \rightarrow CO_2 + H_2O$

    C. $Fe + 3Ag^+ \rightarrow Fe^{3+} + 3Ag$

    D. $2HNO_3 + SO_2 \rightarrow H_2SO_4 + 2NO_2$

    E. None of the above

143. After balancing the following redox reaction, what is the coefficient of CO_2?

$$Co_2O_3\ (s) + CO\ (g) \rightarrow Co\ (s) + CO_2\ (g)$$

    A. 1   **B.** 2   **C.** 3   **D.** 5   **E.** None of the above

144. Round the following number to 3 significant figures: 565.85 grams:

    A. 565.9   **B.** 560   **C.** 565   **D.** 566   **E.** 565.90

145. Oxidation is defined as the:

    A. gaining of protons

    B. losing of protons

    C. gaining of electrons

    D. gaining of neutrons

    E. losing of electrons

Copyright © 2015 Sterling Test Prep

146. How many potassium atoms are in a sample of $K_3Fe(CN)_6$ that contains 1.084×10^{24} carbon atoms?

    A. 1.124×10^{24} atoms   **C.** 2.238×10^{24} atoms

    B. 5.420×10^{23} atoms   **D.** 3.865×10^{23} atoms   **E.** 3.362×10^{24} atoms

147. In which of the following compounds is the oxidation number of hydrogen NOT +1?

    A. NH_3 **B.** $HClO_2$ **C.** H_2SO_4 **D.** NaH   **E.** None of the above

148. What is the term for the volume occupied by 1 mol of any gas at STP?

    A. STP volume   **C.** standard volume

    B. molar volume   **D.** Avogadro's volume   **E.** none of the above

149. $2AgNO_3\ (aq) + K_2SO_4\ (aq) \rightarrow 2KNO_3\ (aq) + Ag_2SO_4\ (s)$.

The net ionic reaction for the balanced equation shown above is:

    A. $2Ag^+ + SO_4^{2-} \rightarrow Ag_2SO_4$ **C.** $2K^+ + SO_4^{2-} \rightarrow K_2SO_4$

    B. $K^+ + NO_3^- \rightarrow KNO_3$ **D.** $Ag^+ + NO_3^- \rightarrow AgNO_3$

    E. $H^+ + OH^- \rightarrow H_2O$

150. If the partial pressure of CO_2 is 60 torr at STP, and the gas is 20% CO_2 by mass, with only one other species of gas, which of the following could be the other species of gas?

    A. ethanol   **B.** sulfur   **C.** methanc   **D.** hydrogen   **E.** bromine

151. How many grams of H_2O can be formed from a reaction between 10 grams of oxygen and 1 gram of hydrogen?

    A. 11 grams of H_2O are formed since mass must be conserved

    B. 10 grams of H_2O are formed since the mass of water produced cannot be greater than the amount of oxygen reacting

    C. 9 grams of H_2O are formed because oxygen and hydrogen react in an 8:1 ratio

    D. No H_2O is formed because there is insufficient hydrogen to react with the oxygen

    E. Not enough information is provided

152. The oxidation number of iron in the compound $FeBr_3$ is:

    A. –2   **B.** +1   **C.** +2   **D.** +3   **E.** –1

153. Which of the following is a method for balancing a redox equation in acidic solution by the half-reaction method?

    A. Multiply each half-reaction by a whole number, so that the number of electrons lost by the substance oxidized is equal to the electrons gained by the substance reduced

    B. Add the two half-reactions together and cancel identical species from each side of the equation

    C. Write a half-reaction for the substance oxidized and the substance reduced

    D. Balance the atoms in each half-reaction; balance oxygen with water and hydrogen with H^+

    E. All of the above

154. Which of the following substances has the lowest density?

    A. A mass of 750 g and a volume of 70 dL   **C.** A mass of 1.5 kg and a volume of 1.2 L

    B. A mass of 5 mg and a volume of 25 μL   **D.** A mass of 25 g and a volume of 20 mL

    E. A mass of 15 mg and a volume of 50 μL

155. How many moles of Sn^{4+} are produced from one mole of Sn^{2+} and excess O_2 and H^+ in the following reaction: $Sn^{2+} + O_2 \rightarrow Sn^{4+}$

    A. 2 moles   **B.** 2.5 moles   **C.** 1 ½ moles   **D.** ½ mole   **E.** 1 mole

156. How many grams of carbon are required to combine with 1.99 g of oxygen in C_2H_5OH?

    A. 1.93 g   **B.** 2.99 g   **C.** 5.26 g   **D.** 4.99 g   **E.** 3.99 g

157. In a redox reaction, the substance that is reduced:

    A. always gains electrons   **C.** contains an element that increases in oxidation number

    B. always loses electrons   **D.** is also the reducing agent

    E. either gains or loses electrons

158. In order for a redox reaction to be balanced, which of the following is true?

    I. Total ionic charge of reactants must equal total ionic charge of products

    II. Atoms of each reactant must equal atoms of product

    III. Electron gain must equal electron loss

    A. I and II only   **C.** II and III only

    B. I and III only   **D.** I, II and III   **E.** None of the above

159. When balanced, the coefficient for CO_2 is:

$$\_\_\_C_5H_{12} + \_\_\_O_2 \rightarrow \_\_\_CO_2 + \_\_\_H_2O$$

    A. 5   **B.** 6   **C.** 8   **D.** 10   **E.** 12

160. How many moles of O_2 gas are required for combustion with 2 moles of hexane in the unbalanced reaction: $C_6H_{14}(g) + O_2(g) \rightarrow CO_2(g) + H_2O(g)$

    A. 11   **B.** 14   **C.** 12   **D.** 20   **E.** 19

161. How many grams of H_2O can be produced from the reaction of 25.0 grams of H_2 and 225 grams of O_2?

    A. 250 grams   **C.** 200 grams

    B. 225 grams   **D.** 25 grams   **E.** 2.50 grams

162. Which of the following represents the oxidation of Co^{2+}?

    A. $Co \rightarrow Co^{2+} + 2e^-$ **C.** $Co^{2+} + 2e^- \rightarrow Co$

    B. $Co^{3+} + e^- \rightarrow Co^{2+}$ **D.** $Co^{2+} \rightarrow Co^{3+} + e^-$ **E.** $Co^{3+} + 2e^- \rightarrow Co^{1+}$

Copyright © 2015 Sterling Test Prep

163. What is the term for the chemical formula of a compound that expresses the actual number of atoms of each element in a molecule?

    A. molecular formula   **C.** elemental formula

    B. empirical formula   **D.** atomic formula   **E.** none of the above

164. What is the coefficient of $HClO_4$ when balanced with smallest whole numbers?

    $Cl_2O_7 + H_2O \rightarrow HClO_4$

    A. 1   **B.** 2   **C.** 3   **D.** 4   **E.** 7

165. A reducing agent:

    A. is oxidized and gains electrons   **C.** is reduced

    B. gains electrons   **D.** loses protons   **E.** is oxidized

166. What is the molecular formula of a compound that has an empirical formula of CHCl with a molar mass of 194g?

    A. $C_4H_4Cl_4$ **B.** $C_2H_2Cl_2$ **C.** $C_3H_3Cl_3$ **D.** CHCl   **E.** $C_4H_3Cl_4$

167. Determine the oxidation number of C in $NaHCO_3$:

    A. +5   **B.** +4   **C.** +12   **D.** +6   **E.** +5

168. After balancing the following redox reaction in acidic solution, what is the coefficient of H^+?

    $Ag\ (s) + NO_3^-\ (aq) \rightarrow Ag^+\ (aq) + NO\ (aq)$

    A. 1   **B.** 2   **C.** 4   **D.** 6   **E.** None of the above

169. When the reaction shown is correctly balanced, the coefficients are:

    $C_6H_{14}\ (l) + O_2\ (g) \rightarrow CO_2\ (g) + H_2O\ (g)$

    A. 1, 3.5, 6, 7   **C.** 1, 6, 6, 7

    B. 1, 9.5, 6, 7   **D.** 2, 19, 12, 14   **E.** 2, 16.5, 12, 7

170. Ten moles of $N_2O_4\ (l)$ are added to an unknown amount of $N_2H_3(CH_3)\ (l)$. When the reaction runs to completion, what is the limiting reagent if 23 moles of H_2O are produced?

    $5N_2O_4\ (l) + 4N_2H_3(CH_3)\ (l) \rightarrow 12H_2O\ (g) + N_2\ (g) + CO_2\ (g)$

    A. $N_2O_4\ (l)$ **C.** $4N_2H_3(CH_3)\ (l)$

    B. $H_2O\ (g)$ **D.** $N_2O_4\ (l)$ **E.** There was no limiting reagent

171. How is Avogadro's number related to the numbers on the periodic table?

    A. The atomic mass listed is the mass of Avogadro's number's worth of atoms

    B. The periodic table provides the mass of one atom, while Avogadro's number provides the number of moles

    C. The masses are all divisible by Avogadro's number, which provides the weight of one mole

    D. The periodic table only provides atomic numbers, not atomic mass

    E. The mass listed is in the units of Avogadro's number

Copyright © 2015 Sterling Test Prep

172. What volume of a 0.400 M $Fe(NO_3)_3$ solution is needed to supply 0.850 moles of nitrate ions?

   A. 345 mL   **B.** 0.708 mL   **C.** 222 mL   **D.** 708 mL   **E.** 984 mL

173. After balancing the following redox reaction in acidic solution, what is the coefficient of H^+?

$$Mg\ (s)\ + NO_3^-\ (aq) \rightarrow Mg^{2+}\ (aq) + NO_2\ (aq)$$

   A. 1   **B.** 2   **C.** 4   **D.** 6   **E.** None of the above

174. How many significant figures are there in the following number: 0.00368 grams?

   A. 1   **B.** 4   **C.** 5   **D.** 2   **E.** 3

175. From the following reaction, if 0.1 mole of Al is allowed to react with 0.2 mole of Fe_2O_3, how many grams of aluminum oxide are produced: $2Al + Fe_2O_3 \rightarrow 2Fe + Al_2O_3$

   A. 3.8 g   **B.** 4.7 g   **C.** 5.1 g   **D.** 8.2 g   **E.** 12.2 g

176. How many atoms are in a sample of phosphorus trifluoride, PF_3, that contains 1.400 moles?

   A. 3.372×10^{24} **C.** 3.4

   B. 5.3   **D.** 2.218×10^{24} **E.** 8.976×10^{23}

177. What is the oxidation number of Br in $NaBrO_4$?

   A. –6   **B.** –4   **C.** +7   **D.** +4   **E.** +5

178. In basic solution, which of the following are guidelines for balancing a redox equation by the half–reaction method?

   I. Verify that the total number of atoms and the total ionic charge are equal for reactants and products

   II. Add the two half–reactions together and cancel identical species on each side of the equation

   III. Multiply each half–reaction by a whole number so that the number of electrons lost by the substance oxidized is equal to the electrons gained by the substance reduced

   IV. Write a balanced half-reaction for the substance oxidized and the substance reduced

   A. I and II only   **C.** II, III and IV only

   B. II and IV only   **D.** I, II and IV only   **E.** I, II, III and IV

179. What is the spectator ion: $2AgNO_3\ (aq) + K_2SO_4\ (aq) \rightarrow 2KNO_3\ (aq) + Ag_2SO_4\ (s)$?

   A. silver ion and sulfate ion   **C.** potassium ion and sulfate ion

   B. potassium ion and nitrate ion   **D.** silver ion and nitrate ion

                                     E. hydrogen ion and hydroxide ion

180. How many moles of O_2 gas are required for combustion with one mole of $C_{12}H_{22}O_{11}$ in the unbalanced reaction: $C_{12}H_{22}O_{11}\ (l) + O_2\ (g) \rightarrow CO_2\ (g) + H_2O\ (g)$

   A. 4   **B.** 8   **C.** 12   **D.** 14   **E.** 20

                                                 Copyright © 2015 Sterling Test Prep

181. What is the formula mass of a molecule of $C_6H_{12}O_6$?

    A. 148 amu   **B.** 27 amu   **C.** 21 amu   **D.** 180 amu   **E.** none of the above

182. For which of the following compounds is the oxidation number of oxygen NOT –2?

    A. $NaClO_2$ **C.** $Ba(OH)_2$

    B. Li_2O_2 **D.** Na_2SO_4 **E.** None of the above

183. Which substance listed below is the strongest reducing agent given the following spontaneous redox reaction?

$$Mg\ (s) + Sn^{2+}\ (aq) \rightarrow Mg^{2+}\ (aq) + Sn\ (s)$$

    A. Sn   **B.** Mg^{2+} **C.** Sn^{2+} **D.** Mg   **E.** none of the above

184. How many H atoms are in the molecule $C_6H_3(C_3H_7)_2(C_2H_5)$?

    A. 27   **B.** 15   **C.** 29   **D.** 10   **E.** 22

185. In the following reaction: $Cu^{2+} + 2e^- \rightarrow Cu\ (s)$

    A. Cu^{2+} is neutralized   **C.** Cu^{2+} is oxidized

    B. Cu^{2+} is reduced and neutralized   **D.** Cu^{2+} is reduced   **E.** Cu^{2+} is a reducing agent

186. Upon combustion analysis, a 6.987 g sample of a hydrocarbon yielded 8.398 grams of carbon dioxide. The percent, by mass, of carbon in the hydrocarbon is:

    A. 18.26 %   **B.** 23.27 %   **C.** 32.80 %   **D.** 31.37 %   **E.** 52.28 %

187. Which equation is NOT correctly classified by the type of chemical reaction?

Equation	Reaction Type
A. $AgNO_3 + NaCl \rightarrow AgCl + NaNO_3$	double-displacement/non-redox
B. $Cl_2 + F_2 \rightarrow 2\ ClF$	synthesis/redox
C. $H_2O + SO_2 \rightarrow H_2SO_3$	synthesis/non-redox
D. $CaCO_3 \rightarrow CaO + CO_2$	decomposition/redox
E. All are correctly classified	

188. Which substance is the weakest reducing agent given the spontaneous redox reaction?

$$Mg\ (s) + Sn^{2+}\ (aq) \rightarrow Mg^{2+}\ (aq) + Sn\ (s)$$

    A. Sn   **B.** Mg^{2+} **C.** Sn^{2+} **D.** Mg   **E.** None of the above

189. The balanced equation for the reaction occurring when iron (III) oxide, a solid, is reduced with pure carbon to produce carbon dioxide and molten iron is:

    A. $2Fe_2O_3 + 3C\ (s) \rightarrow 4Fe\ (l) + 3CO_2\ (g)$

    B. $4Fe_2O_3 + 6C\ (s) \rightarrow 8Fe\ (l) + 6CO_2\ (g)$

    C. $2FeO_3 + 3C\ (s) \rightarrow 2Fe\ (l) + 3CO_2\ (g)$

    D. $2Fe_3O + C\ (s) \rightarrow 6Fe\ (l) + CO_2\ (g)$

    E. $2FeO + C\ (s) \rightarrow 2Fe\ (l) + CO_2\ (g)$

190. 13.5 moles of N_2 gas are mixed with 33 moles of H_2 gas in the following reaction:

$$N_2(g) + 3H_2(g) \rightarrow 2NH_3(g)$$

How many moles of N_2 gas remain if the reaction produces 18 moles of NH_3 gas when performed at 600 K?

    A. 0 moles   **B.** 2.25 moles   **C.** 4.5 moles   **D.** 9.0 moles   **E.** 18.0 moles

191. Which is a correctly balanced equation?

    A. $2P_4 + 12H_2 \rightarrow 8PH_3$ **C.** $P_4 + 6H_2 \rightarrow 4PH_3$

    B. $P_4 + 6H_2 \rightarrow 4PH_3$ **D.** $P_4 + 6H_2 \rightarrow 4PH_3$ **E.** $P_4 + 3H_2 \rightarrow PH_3$

192. Which element is reduced in the redox reaction: $BaSO_4 + 4C \rightarrow BaS + 4CO$?

    A. O in CO   **B.** Ba in BaS   **C.** C   **D.** C in CO   **E.** S in $BaSO_4$

193. What is the term for the amount of substance that contains 6.02×10^{23} particles?

    A. molar mass   **C.** Avogadro's number

    B. mole   **D.** formula mass   **E.** none of the above

194. What is the coefficient for O_2 when balanced with the lowest whole number coefficients?

$$C_2H_6 + O_2 \rightarrow CO_2 + H_2O$$

    A. 3   **B.** 4   **C.** 6   **D.** 7   **E.** 9

195. What is the oxidation number of Cr in the compound HCr_2O_4Cl?

    A. +3   **B.** +2   **C.** +6   **D.** +5   **E.** +4

196. Select the balanced chemical equation for the reaction: $C_6H_{14} + O_2 \rightarrow CO_2 + H_2O$

    A. $3C_6H_{14} + O_2 \rightarrow 18CO_2 + 22H_2O$ **C.** $2C_6H_{14} + 19O_2 \rightarrow 12CO_2 + 14H_2O$

    B. $2C_6H_{14} + 12O_2 \rightarrow 12CO_2 + 14H_2O$ **D.** $2C_6H_{14} + 9O_2 \rightarrow 12CO_2 + 7H_2O$

    E. $C_6H_{14} + O_2 \rightarrow CO_2 + H_2O$

197. Which of the following reactions is NOT correctly classified?

    A. $AgNO_3(aq) + KOH(aq) \rightarrow KNO_3(aq) + AgOH(s)$ non-redox / precipitation

    B. $2H_2O_2(s) \rightarrow 2H_2O(l) + O_2(g)$ non-redox / decomposition

    C. $Pb(NO_3)_2(aq) + 2Na(s) \rightarrow Pb(s) + 2NaNO_3(aq)$ redox / single-replacement

    D. $HNO_3(aq) + LiOH(aq) \rightarrow LiNO_3(aq) + H_2O(l)$ non-redox / double-displacement

    E. All are correctly classified

198. Which substance listed below is the weakest oxidizing agent given the following *spontaneous* redox reaction?

$$FeCl_3(aq) + NaI(aq) \rightarrow I_2(s) + FeCl_2(aq) + NaCl(aq)$$

    A. I_2 **B.** $FeCl_2$ **C.** $FeCl_3$ **D.** NaI   **E.** NaCl

    Copyright © 2015 Sterling Test Prep

199. The oxidation number of iron in the compound $FeBr_3$ is:

   A. –2   **B.** –1   **C.** +1   **D.** +2   **E.** +3

200. Which could NOT be true for the following reaction: $N_2(g) + 3H_2(g) \rightarrow 2NH_3(g)$?

   A. 25 grams of N_2 gas reacts with 75 grams of H_2 gas to form 50 grams of NH_3 gas
   B. 28 grams of N_2 gas reacts with 6 grams of H_2 gas to form 34 grams of NH_3 gas
   C. 15 moles of N_2 gas reacts with 45 moles of H_2 gas to form 30 moles of NH_3 gas
   D. 5 molecules of N_2 gas reacts with 15 molecules of H_2 gas to form 10 molecules of NH_3 gas
   E. None of the above

201. Which coefficients balance the following equation: ___$P_4(s)$ + ___$H_2(g) \rightarrow$ ___$PH_3(g)$?

   A. 2, 10, 8   **B.** 1, 4, 4   **C.** 1, 6, 4   **D.** 4, 2, 3   **E.** 1, 3, 4

202. Which response represents the balanced half reaction for *oxidation* for the reaction:

   $HCl(aq) + Fe(s) \rightarrow FeCl_3(aq) + H_2(g)$

   A. $2Fe \rightarrow 2Fe^{3+} + 6e^-$ **C.** $Fe + 3e^- \rightarrow Fe^{3+}$
   B. $3Fe \rightarrow Fe^{3+} + 3e^-$ **D.** $6e^- + 6H^+ \rightarrow 3H_2$ **E.** None of the above

203. From the periodic table, how many atoms of cobalt equal a mass of 58.93 g?

   A. 58.93   **B.** 59   **C.** 1   **D.** 6.02×10^{23} **E.** 29.5

204. What is the total of all the coefficients when balanced with the lowest whole number coefficients?

   $N_2H_4 + H_2O_2 \rightarrow N_2 + H_2O$

   A. 4   **B.** 8   **C.** 10   **D.** 12   **E.** 14

205. Under acidic conditions, what is the sum of the coefficients in the balanced reactions below?

   $Fe^{2+} + Cr_2O_7^{2-} \rightarrow Fe^{3+} + Cr^{3+}$

   A. 8   **B.** 14   **C.** 17   **D.** 36   **E.** None of the above

206. The oxidation states of sulfur in H_2SO_5 and H_2SO_3, respectively, are:

   A. +2 and +6   **C.** +2 and +4
   B. +6 and +4   **D.** +4 and +2   **E.** +4 and +4

207. Which substance is functioning as the reducing agent:

   $14H^+ + Cr_2O_7^{2-} + 3\,Ni \rightarrow 3Ni^{2+} + 2Cr^{3+} + 7H_2O$

   A. $Cr_2O_7^{2-}$ **B.** H_2O **C.** Ni   **D.** H^+ **E.** Ni^{2+}

208. What is the term for a chemical formula that expresses the simplest whole number ratio of atoms of each element in a molecule?

   A. empirical formula   **C.** atomic formula
   B. molecular formula   **D.** elemental formula   **E.** none of the above

209. The balanced equation for the reaction occurring when calcium nitrate solution is mixed with sodium phosphate solution is:

    A. $3CaNO_3$ (aq) + Na_3PO_4 (aq) → Ca_3PO_4 (aq) + $3NaNO_3$ (s)

    B. $Ca(NO_3)_2$ (aq) + $2NaPO_4$ (aq) → $Ca(PO_4)_2$ (s) + $2NaNO_3$ (aq)

    C. $3Ca(NO_3)_2$ (aq) + $2Na_3PO_4$ (aq) → $Ca_3(PO_4)_2$ (aq) + $6NaNO_3$ (aq)

    D. $2Ca(NO_3)_2$ (aq) + $3Na_3PO_4$ (aq) → $2 Ca_3(PO_4)_2$ (s) + $6NaNO_3$ (aq)

    E. $3Ca(NO_3)_2$ (aq) + $2Na_3PO_4$ (aq) → $Ca_3(PO_4)_2$ (s) + $6NaNO_3$ (aq)

210. How much phosphorous is required to produce 275 g of phosphorous trichloride when the following reaction is run to completion: P_4 (s) + $6Cl_2$ (g) → PCl_3 (l)

    A. 62 g   **B.** 124 g   **C.** 166 g   **D.** 238 g   **E.** 265 g

211. Balance the equation: ___H_2 (g) + ___N_2 (g) → ___NH_3 (g)

    A. 3, 2, 2   **B.** 3, 1, 2   **C.** 2, 2, 3   **D.** 2, 2, 5   **E.** 3, 1, 2

212. Which equation is NOT correctly classified by the type of chemical reaction?

Equation	Reaction Type
A. $PbO + C → Pb + CO$	single-replacement/non-redox
B. $2Na + 2HCl → 2NaCl + H_2$	single-replacement/redox
C. $NaHCO_3 + HCl → NaCl + H_2O + CO_2$	double-replacement/non-redox
D. $2Na + H_2 → 2NaH$	synthesis/redox
E. All are correctly classified	

213. Which substance is the weakest reducing agent given the spontaneous redox reaction?

    $FeCl_3$ (aq) + NaI (aq) → I_2 (s) + $FeCl_2$ (aq) + $NaCl$ (aq)

    A. NaCl   **B.** I_2 **C.** NaI   **D.** $FeCl_3$ **E.** $FeCl_2$

214. How many O atoms are in the formula unit $GaO(NO_3)_2$?

    A. 3   **B.** 4   **C.** 5   **D.** 7   **E.** 8

215. What is the oxidation state of S in sulfuric acid?

    A. +8   **B.** +6   **C.** –2   **D.** –6   **E.** +4

216. How many electrons are lost or gained by each formula unit of $CuBr_2$ in this reaction?

    $Zn + CuBr_2 → ZnBr_2 + Cu$

    A. loses 1 electron   **C.** gains 2 electrons

    B. gains 6 electrons   **D.** loses 2 electrons   **E.** gains 4 electrons

217. What general term refers to the mass of 1 mol of any substance?

    A. gram-formula mass   **C.** gram-atomic mass

    B. molar mass   **D.** gram-molecular mass   **E.** none of the above

    Copyright © 2015 Sterling Test Prep

218. The net ionic equation for the reaction between zinc and hydrochloric acid solution is:

A. $Zn^{2+} (aq) + H_2 (g) \rightarrow ZnS (s) + 2H^+ (aq)$

B. $ZnCl_2 (aq) + H_2 (g) \rightarrow ZnS (s) + 2HCl (aq)$

C. $Zn (s) + 2H^+ (aq) \rightarrow Zn^{2+} (aq) + H_2 (g)$

D. $Zn (s) + 2HCl (aq) \rightarrow ZnCl_2 (aq) + H_2 (g)$

E. None of these

219. If the partial pressure of CO_2 is 60 torr at STP, and nitrogen is the only other gas present, what is the percent by mass of CO_2?

A. 1%   B. 25%   C. 40.5%   D. 3%   E. 12%

220. For a basic solution, which is the correct balanced half reaction?

$$Cr(OH)_4^- \rightarrow CrO_4^{2-}$$

A. $Cr(OH)_4^- \rightarrow CrO_4^{2-} + 4H_2O + 2e^-$

B. $2OH^- + Cr(OH)_4^- \rightarrow CrO_4^{2-} + 6e^- + 3H_2O$

C. $4OH^- + Cr(OH)_4^- \rightarrow CrO_4^{2-} + 3e^- + 4H_2O$

D. $3e^- + Cr(OH)_4^- + 2OH^- \rightarrow CrO_4^{2-} + 2H_2O$

E. None of the above

221. What principle states that equal volumes of gases, at the same temperature and pressure, contain equal numbers of molecules?

A. Dalton's theory

B. Charles' theory

C. Boyle's theory

D. Avogadro's theory   E. None of the above

222. If 64 g of O_2 gas are reacted in the following reaction, how many moles of water are produced?

$$Sn^{2+} + O_2 \rightarrow Sn^{4+}$$

A. 3 moles   B. 2 moles   C. 6 moles   D. 5 moles   E. 4 moles

223. How many moles of phosphorous trichloride are required to produce 365 grams of HCl when the reaction yields 75%: $PCl_3 (g) + 2NH_3 (g) \rightarrow P(NH_2)_3 + 3HCl (g)$.

A. 1 mol   B. 2.5 mol   C. 3.5 mol   D. 4.5 mol   E. 5 mol

224. Which response represents the balanced half reaction for *reduction* for the reaction given below?

$$Fe (s) + CuSO_4 (aq) \rightarrow Fe_2(SO4)_3(aq) + Cu (s)$$

A. $3Cu^{2+} + 6e^- \rightarrow 3Cu$

B. $2Fe \rightarrow 2Fe^{3+} + 6e^-$

C. $2Cu^{3+} + 3e^- \rightarrow 2Cu$

D. $Fe + 3e^- \rightarrow Fe^{3+}$

E. none of the above

225. What is the term for a temperature of 0 °C and a pressure of 1 atm?

A. standard temperature and pressure

B. ideal gas temperature and pressure

C. experimental temperature and pressure

D. atmospheric temperature and pressure

E. none of the above

226. What are the spectator ions in the reaction between KOH and HNO_3?

    A. K^+ and NO_3^- **C.** K^+ and H^+

    B. H^+ and NO_3^- **D.** H^+ and OH^- **E.** K^+ and OH^-

227. A 10 Newton weight is placed on a 6-centimeter radius piston, which compresses He in a sealed container until the piston stops moving because of the increase in pressure. Some He is then removed causing the piston to fall, so that the container's volume decreases by 25% and the gas again supports the piston. How do the two pressures compare?

    A. There is not enough information to answer the question

    B. The pressures are the same because the force on the piston is constant

    C. The second pressure is higher because the pressure and volume are inversely proportional

    D. The second pressure is lower because pressure and volume are directly proportional

    E. The second pressure is lower because pressure and volume are inversely proportional

228. If the relative mass of a ping pong ball is 1/20 that of a golf ball, how many ping pong balls are needed to equal the mass of two golf balls?

    A. 100   **B.** 10   **C.** 20   **D.** 40   **E.** 6.022×10^{23}

229. In the disproportion equation below, what species is undergoing disproportionation?

$$3Br_2 + 6OH^- \rightarrow BrO_3^- + 5Br^- + 3H_2O$$

    A. Br_2 **B.** H_2O **C.** Br^- **D.** OH^- **E.** None of the above

230. What is the volume occupied by 0.750 g of N_2 gas, at STP?

    A. 37.3 L   **B.** 16.8 L   **C.** 0.600 L   **D.** 0.938 L   **E.** 836 L

231. For *n* moles of gas, which term expresses the kinetic energy?

    A. nPA, where n = number of moles of gas, P = total pressure and A = surface area of the container walls

    B. 1/2 nPA, where n = number of moles of gas, P = total pressure and A = surface area of the container walls

    C. 1/2 MV^2, where M = molar mass of the gas and V = volume of the container

    D. MV^2, where M = molar mass of the gas and V = volume of the container

    E. 3/2 nRT, where n = number of moles of gas, R = ideal gas constant and T = absolute temperature

232. How many grams of Ba^{2+} ions are in an aqueous solution of $BaCl_2$ that contains 6.8×10^{22} Cl ions?

    A. 3.2×10^{48} g   **B.** 12g   **C.** 14.5g   **D.** 7.8g   **E.** 9.8g

233. The formula for mustard gas used in chemical warfare is $C_4H_8SCl_2$ (159.09 g/mol). What is the percentage of chlorine in mustard gas?

    A. 20.16%   **B.** 44.57%   **C.** 5.08%   **D.** 30.20%   **E.** 22.28%

Copyright © 2015 Sterling Test Prep

234. Which is a good experimental method to distinguish between ordinary hydrogen and deuterium, the rare isotope of hydrogen?

       I. Measure the density of the gas at STP

       II. Measure the rate at which the gas effuses

       III. Determine the number of grams of gas that react with one mole of O_2 to form H_2O

A. I only   **B.** II only   **C.** III only   **D.** I, II, and III   **E.** None of the above

235. Which reaction is the correctly balanced half reaction (in acid solution) for the process below?

$$Cr_2O_7^{2-} (aq) \rightarrow Cr^{3+} (aq)$$

A. $8H^+ + Cr_2O_7 \rightarrow 2Cr^{3+} + 4H_2O + 3e^-$

B. $12H^+ + Cr_2O_7^{2-} + 3e^- \rightarrow 2Cr^{3+} + 6H_2O$

C. $14H^+ + Cr_2O_7^{2-} + 6e^- \rightarrow 2Cr^{3+} + 7H_2O$

D. $8H^+ + Cr_2O_7 + 3e^- \rightarrow 2Cr^{3+} + 4H_2O$

E. None of the above

236. In the early 1980s, benzene that had been used as a solvent for waxes and oils was listed as a carcinogen by the EPA. What is the molecular formula of benzene if the empirical formula is C_1H_1, and the approximate molar mass is 78 g/mol?

A. CH_{12} **B.** $C_{12}H_{12}$ **C.** CH   **D.** CH_6 **E.** C_6H_6

237. Under ideal conditions, which of the following gases is least likely to be ideal?

A. CCl_4 **B.** CH_3OH **C.** O_2 **D.** O_3 **E.** CO_2

238. The reason why ice floats in a glass of water is because, when frozen, H_2O is less dense due to:

A. strengthening of cohesive forces   **D.** weakening of cohesive forces

B. high specific heat   **E.** increased number of hydrogen bonds

C. decreased number of hydrogen bonds

239. What is the molecular formula of galactose if the empirical formula is CH_2O, and the approximate molar mass is 180 g/mol?

A. $C_6H_{12}O_6$ **B.** CH_2O_6 **C.** CH_2O **D.** CHO   **E.** $C_{12}H_{22}O_{11}$

240. As the volume of a sealed container is decreased at constant temperature, the gas begins to deviate from ideal behavior. Compared to the pressure predicted by the ideal gas law, the actual pressure is:

A. higher because of intermolecular attractions between gas molecules

B. higher because of the volume of the gas molecules

C. lower because of intermolecular attractions among gas molecules

D. lower because of the volume of the gas molecules

E. higher because of intramolecular attractions between gas molecules

Copyright © 2015 Sterling Test Prep

Chapter 7.   EQUILIBRIUM & REACTION RATES

1. What is the term for the energy necessary for reactants to achieve the transition state and form products?

    A. heat of reaction   **C.** rate barrier

    B. energy barrier   **D.** collision energy   **E.** activation energy

2. Which of the following stresses would shift the equilibrium to the left for the following chemical system at equilibrium *heat* + $6H_2O$ (g) + $2N_2$ (g) $\leftrightarrow$ $4NH_3$ (g) + $3O_2$ (g)?

    A. Increasing the concentration of H_2O **C.** Increasing the concentration of O_2

    B. Decreasing the concentration of NH_3 **D.** Increasing the reaction temperature

                                       E. Decreasing the concentration of O_2

3. Which component of the reaction mechanism results in the net production of free radicals?

    I. initiation   II. propagation   III. termination

    A. I only   **B.** II only   **C.** III only   **D.** II and III only   **E.** I and III only

4. Nitrogen monoxide reacts with bromine at elevated temperatures according to the equation

$$2NO\ (g) + Br_2\ (g) \rightarrow 2NOBr\ (g)$$

What is the rate of consumption of Br_2 (g) if, in a certain reaction mixture, the rate of formation of NOBr (g) was 4.50×10^{-4} mol L^{-1} s^{-1}?

    A. 3.12×10^{-4} mol L^{-1} s^{-1} **C.** 2.25×10^{-4} mol L^{-1} s^{-1}

    B. 8.00×10^{-4} mol L^{-1} s^{-1} **D.** 4.50×10^{-4} mol L^{-1} s^{-1}

                                       E. 4.00×10^{-3} mol L^{-1} s^{-1}

5. For a collision between molecules to result in a reaction, the molecules must possess both a favorable orientation relative to each other and:

    A. be in the gaseous state   **C.** adhere for at least 2 nanoseconds

    B. have a certain minimum energy   **D.** exchange electrons

                                       E. be in the liquid state

6. What is the term for a substance that allows a reaction to proceed faster by lowering the energy of activation?

    A. rate barrier   **C.** collision energy

    B. energy barrier   **D.** activation energy   **E.** catalyst

7. Reaction rates are determined by all of the following factors, EXCEPT:

    A. orientation of collisions between molecules

    B. spontaneity of the reaction

    C. force of collisions between molecules

    D. number of collisions between molecules

    E. the activation energy of the reaction

Copyright © 2015 Sterling Test Prep

8. For the combustion of ethanol (C_2H_6O) to form carbon dioxide and water, what is the rate at which carbon dioxide is produced if the ethanol is consumed at a rate of 3.0 M s^{-1}?

A. 1.5 M s^{-1}

B. 12.0 M s^{-1}

C. 6.0 M s^{-1}

D. 9.0 M s^{-1}

E. 10.0 M s^{-1}

9. If the heat of reaction is endothermic, which of the following is always true?

A. The energy of the reactants is less than the products

B. The energy of the reactants is greater than the products

C. the reaction rate is slow

D. The reaction rate is fast

E. None of the above

10. Most reactions are carried out in liquid solution or in the gaseous phase, because in such situations:

A. kinetic energies of reactants are lower

B. reactant collisions occur more frequently

C. activation energies are higher

D. reactant activation energies are lower

E. reactant collisions occur less frequently

11. Why might increasing the concentration of a set of reactants increase the rate of reaction?

A. The rate of reaction depends only on the mass of the atoms and therefore increases as you increase the mass of the reactants

B. The concentration of reactants is unrelated to the rate of reaction

C. There is an increased ratio of reactants to products

D. There is an increased probability that any two reactant molecules collide and react

E. None of the above

12. What is the term for a dynamic state of a reversible reaction in which the rates of the forward and reverse reactions are equal?

A. dynamic equilibrium

B. rate equilibrium

C. reversible equilibrium

D. concentration equilibrium

E. chemical equilibrium

13. Which conditions would favor the reaction to completion?

$2N_2 (g) + 6H_2O (g) + heat \leftrightarrow 4NH_3 (g) + 3O_2 (g)$

A. Increase reaction temperature

B. Continual addition of NH_3 gas to the reaction mixture

C. Decrease the pressure on the reaction vessel

D. Continual removal of N_2 gas

E. Decreases reaction temperature

14. The reaction $\cdot OH + H_2 \rightarrow H_2O + H\cdot$ is an example of a free radical:

I. initiation   II. propagation   III. termination

A. I only   **B.** II only   **C.** III only   **D.** II and III only   **E.** I and II only

Copyright © 2015 Sterling Test Prep

15. In a particular study of the reaction described by the equation, what is the rate of consumption of CH_4O (g) if the rate of consumption of $O_2(g)$ is 0.400 mol $L^{-1}s^{-1}$?

$$2CH_4O\,(g) + 3O_2\,(g) \rightarrow 2CO_2\,(g) + 4H_2O\,(g)$$

A. 0.150 mol $L^{-1}s^{-1}$

B. 0.267 mol $L^{-1}s^1$

C. 0.333 mol $L^{-1}s^{-1}$

D. 0.550 mol $L^{-1}s^{-1}$

E. 0.500 mol $L^{-1}s^{-1}$

16. Which of the following statements about "activation energy" is correct?

A. Activation energy is the energy given off when reactants collide

B. Activation energy is high for reactions that occur rapidly

C. Activation energy is low for reactions that occur rapidly

D. Activation energy is the maximum energy a reacting molecule may possess

E. Activation energy is low for reactions that occur slowly

17. What is the term for the principle that the rate of reaction is regulated by the frequency, energy and orientation of molecules striking each other?

A. orientation theory

B. frequency theory

C. energy theory

D. collision theory

E. rate theory

18. Which of the following statements can be assumed to be true about how reactions occur?

A. Reactant particles must collide with each other

B. Energy must be released as the reaction proceeds

C. Catalysts must be present in the reaction

D. Energy must be absorbed as the reaction proceeds

E. The energy of activation must have a negative value

19. Given the rate = $k[A]^2[B]^4$, what is the order of the reaction?

A. 1   B. 2   C. 4   D. 6   E. 8

20. What is the equilibrium constant expression (K_{sp}) for slightly soluble silver sulfate in an aqueous solution $Ag_2SO_4\,(s) \rightleftarrows 2Ag^+\,(aq) + SO_4^{2-}\,(aq)$?

A. $K_{sp} = [Ag^+]\,[SO_4^{2-}]^2$

B. $K_{sp} = [Ag^+]^2\,[SO_4^{2-}]\,/\,[Ag_2SO_4]$

C. $K_{sp} = [Ag^+]\,[SO_4^{2-}]$

D. $K_{sp} = [Ag^+]\,[SO_4^{2-}]^2\,/\,[Ag_2SO_4]$

E. $K_{sp} = [Ag^+]^2\,[SO_4^{2-}]$

21. The minimum combined kinetic energy reactants must possess for collisions to result in a reaction is:

A. orientation energy

B. activation energy

C. collision energy

D. dissociation energy

E. bond energy

22. Why does a glowing splint of wood burn only slowly in air, but rapidly in a burst of flames when placed in pure oxygen?

    A. A glowing wood splint is actually extinguished within pure oxygen because oxygen inhibits the smoke

    B. Pure oxygen is able to absorb carbon dioxide at a faster rate

    C. Oxygen is a flammable gas

    D. There is an increased number of collisions between the wood and oxygen molecules

    E. There is a decreased number of collisions between the wood and oxygen

23. What is the term for a reaction that proceeds by absorbing heat energy?

    A. isothermal reaction

    B. exothermic reaction

    C. endothermic reaction

    D. all of the above

    E. none of the above

24. All of the following factors may shift the position of equilibrium, EXCEPT:

    A. decreasing reaction temperature

    B. doubling the pressure

    C. increasing reaction temperature

    D. reducing reaction volume

    E. addition of a catalyst

25. Which factor does NOT describe activated complexes?

    A. May be chemically isolated

    B. Decompose rapidly

    C. Have specific geometry

    D. Are extremely reactive

    E. At the high point of the reaction profile

26. A 10-mm cube of copper metal is placed in 250 mL of 12 M nitric acid at 25 °C and the reaction below occurs:

$$Cu(s) + 4H^+ (aq) + 2NO_3^- (aq) \rightarrow Cu^{2+} (aq) + 2NO_2 (g) + 2H_2O (l)$$

At a particular instant in time, nitrogen dioxide is being produced at the rate of 2.6×10^{-4} M/min. At this same instant, what is the rate at which hydrogen ions are being consumed?

    A. 2.6×10^{-4} M/min

    B. 1.0×10^{-3} M/min

    C. 1.3×10^{-4} M/min

    D. 5.2×10^{-4} M/min

    E. 6.5×10^{-5} M/min

27. If, at equilibrium, reactant concentrations are slightly smaller than product concentrations, the equilibrium constant would be:

    A. slightly greater than 1

    B. slightly lower than 1

    C. $\ll 1$

    D. $\gg 1$

    E. equal to zero

28. Which of the following influences the rate of a chemical reaction?

    A. Collision orientation

    B. Collision energy

    C. Collision frequency

    D. None of the above

    E. All of the above

Copyright © 2015 Sterling Test Prep

29. Which of the following affects all reaction rates?

    A. Temperature of the reactants

    B. Presence of a catalyst

    C. Concentrations of the reactants

    D. All are correct

    E. None affect the rates

30. If the rate law $= k[A]^{1/2}[B]$, which of the following can be concluded about the reaction?

    A. Reaction mechanism does not occur in a single step

    B. The rate of A is half the rate of B

    C. For every molecule of B, two molecules of A react

    D. For every molecule of A, two molecules of B react

    E. None are true

31. Which of the changes has no effect on the equilibrium for the reversible reaction:

    $$SO_3\,(g) + NO\,(g) + heat \leftrightarrow SO_2\,(g) + NO_2\,(g)$$

    A. add a catalyst

    B. add helium gas

    C. increase volume

    D. decrease volume

    E. all of the above

32. The expression for the equilibrium constant, K_{eq}, for the reaction below is:

    $$4NH_3\,(g) + 5O_2\,(g) \leftrightarrow 4NO\,(g) + 6H_2O\,(g)$$

    A. $K_{eq} = [NO]^4[H_2O]^6 / [NH_3]^4[O_2]^5$

    B. $K_{eq} = [NH_3]^4[O_2]^5 / [NO]^4[H_2O]^6$

    C. $K_{eq} = [NO][H_2O] / [NH_3][O_2]$

    D. $K_{eq} = [NH_3][O_2] / [NO][H_2O]$

    E. $K_{eq} = [NH_3]^2[O_2]^5 / [NO]^2[H_2O]^3$

33. Why might increasing temperature alter the rate of a chemical reaction?

    A. The molecules combine with other atoms at high temperature to save space

    B. The density decreases as a function of temperature that increases volume and decreases reaction rate

    C. The molecules have a higher kinetic energy and have more force when colliding

    D. The molecules are less reactive at higher temperatures

    E. None of the above

34. If the E_{act} is lowered, which of the following is always true?

    A. The reaction is endothermic

    B. The reaction is exothermic

    C. The reaction proceeds slower

    D. The reaction proceeds faster

    E. None of the above

35. If a reaction proceeds in several steps, the process with the highest activation energy is the:

    A. product formation step

    B. activated complex step

    C. transition step

    D. favorable step

    E. rate-determining step

36. What is the overall order of the reaction if the units of the rate constant for a particular reaction are min^{-1}?

    A. Zero   B. First   C. Second   D. Third   E. Fourth

37. Which statement is NOT a correct characterization for a catalyst?

    A. Catalysts are not consumed in a reaction

    B. Catalysts do not actively participate in a reaction

    C. Catalysts lower the activation energy for a reaction

    D. Catalysts can be either solids, liquids or gases

    E. Catalysts do not alter the equilibrium of the reaction

38. What is the term for a reaction that proceeds by releasing heat energy?

    A. Endothermic reaction   **C.** Exothermic reaction

    B. Isothermal reaction   **D.** All of the above   **E.** None of the above

39. For a chemical reaction to occur, all of the following must happen, EXCEPT:

    A. reactant particles must collide with the correct orientation

    B. a large enough number of collisions must occur

    C. chemical bonds in the reactants must break

    D. reactant particles must collide with enough energy for change to occur

    E. chemical bonds in the products must form

40. By convention, what is the equilibrium constant for step 1 in a reaction?

    A. $k_1 k_{+2}$ **B.** $k_1 + k_{+2}$ **C.** k_{-1} **D.** k_1 **E.** k_1/k_{-1}

41. What is the K_{sp} for calcium fluoride (CaF_2) if the calcium ion concentration in a saturated solution is 0.00021 M?

    A. $K_{sp} = 3.7 \times 10^{-11}$ **C.** $K_{sp} = 4.4 \times 10^{-8}$

    B. $K_{sp} = 2.3 \times 10^{-12}$ **D.** $K_{sp} = 8.8 \times 10^{-8}$ **E.** $K_{sp} = 9.3 \times 10^{-12}$

42. Which statement is correct concerning Reaction A if it releases 24 kJ/mole and has an activation energy of 98 kJ/mole?

    A. The overall reaction is exothermic with a high activation energy

    B. The overall reaction is endothermic with a high activation energy

    C. The overall reaction is exothermic with a low activation energy

    D. The overall reaction is endothermic with a low activation energy

    E. None of the above

43. Why does dough rise faster in a warmer area when yeast in bread dough feed on sugar to produce carbon dioxide?

    A. The yeast tends to become activated with warmer temperatures, which is why baker's yeast is best stored in the refrigerator

    B. The rate of evaporation increases with increasing temperature

    C. There is a greater number of effective collisions among reacting molecules

    D. Atmospheric pressure decreases with increasing temperature

    E. Atmospheric pressure increases with increasing temperature

 Copyright © 2015 Sterling Test Prep

44. Which is true before a reaction reaches chemical equilibrium?

   A. The amount of reactants and products are equal

   B. The amount of reactants and products are constant

   C. The amount of products is decreasing

   D. The amount of reactants is increasing

   E. The amount of products is increasing

45. Which of the changes shifts the equilibrium to the right for the following system at equilibrium:

$$N_2 (g) + 3H_2 (g) \leftrightarrow 2NH_3 (g) + 92.94 \text{ kJ}$$

   1. Increasing temperature   2. Decreasing temperature

   3. Increasing volume   4. Decreasing volume

   5. Removing NH_3 6. Adding NH_3

   7. Removing N_2 8. Adding N_2

   A. 2, 4, 5, 8   **B.** 1, 6, 8   **C.** 2, 3, 5, 8   **D.** 1, 4, 5, 8   **E.** 1, 3, 5, 6

46. Which is NOT an important condition for a chemical reaction?

   A. The molecules must make contact

   B. The molecules have enough energy to react once they collide

   C. The reacting molecules are in the correct orientation to one another

   D. None of the above

   E. The molecules are in the solid, liquid or gaseous state

47. The rate of a chemical reaction in solution can be measured in the units:

   A. mol s L^{-1} **C.** $\text{mol L}^{-1} \text{ s}^{-1}$

   B. s^{-2} **D.** $\text{L}^2 \text{ mol}^{-1} \text{ s}^{-1}$ **E.** $\text{sec L}^{-1} \text{ mol}^{-1}$

48. Whether a reaction is exothermic or endothermic is determined by:

   A. an energy balance between bond breaking and bond forming, resulting in a net loss or gain of energy

   B. whether a catalyst is present

   C. the activation energy

   D. the physical state of the reaction system

   E. none of the above

49. Which of the changes shifts the equilibrium to the right for the following reversible reaction?

$$SO_3 (g) + NO (g) + heat \leftrightarrow SO_2 (g) + NO_2 (g)$$

   A. decreasing temperature   **C.** increasing $[SO_3]$

   B. increasing volume   **D.** increasing $[SO_2]$ **E.** adding a catalyst

50. In the reaction energy diagrams shown below, reaction A is [   ] and occurs [   ] reaction B.

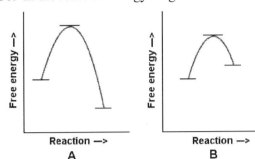

A. endergonic… slower than

 B. exergonic… slower than

C. endergonic… faster than

D. exergonic… faster than

E. exergonic… at the same rate as

51. In the reaction energy diagrams shown in question 50, reaction B is [   ] and occurs [   ] reaction A.

A. exergonic… slower than

B. endergonic… slower than

C. exergonic… faster than

D. exergonic… at the same rate as

E. endergonic… faster than

52. What is the order of the reaction when the rate law = $k[A][B]^3$?

A. 2   **B.** 3   **C.** 4   **D.** 5   **E.** 6

53. What is the equilibrium constant expression, K_i, for the following weak acid?

$$H_3PO_4\,(aq) \leftrightarrow H^+\,(aq) + H_2PO_4^-\,(aq)$$

A. $K_i = [H_3PO_4] / [H^+] [H_2PO_4^-]$

B. $K_i = [H^+]^3 [PO_4^{3-}] / [H_3PO_4]$

C. $K_i = [H^+]^3 [H_2PO_4^-] / [H_3PO_4]$

D. $K_i = [H^+] [H_2PO_4^-] / [H_3PO_4]$

E. $K_i = [H_3PO_4] / [H^+]^3 [PO_4^{3-}]$

54. Which reaction below is endothermic?

A. $PCl_3 + Cl_2 \rightarrow PCl_5 + heat$

B. $2NO_2 \rightarrow N_2 + 2O_2 + heat$

C. $CH_4 + N_2 + heat \rightarrow hn + NH_3$

D. $NH_3 + hr \rightarrow NH_4Br + heat$

E. $PCl_3 + Cl_2 \rightarrow PCl_5 + heat$

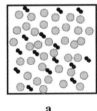 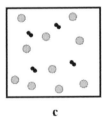

a   b   c

$$X + Y \rightarrow Z$$

55. Which of the reactions proceeds the fastest?

A. a   **C.** c

B. b   **D.** all proceed at same rate   **E.** not enough information given

56. Which of the reactions proceeds the slowest?

A. a   **C.** c

B. b   **D.** all proceed at same rate   **E.** not enough information given

     Copyright © 2015 Sterling Test Prep

57. What equilibrium constant applies to a reversible reaction involving a gaseous mixture at equilibrium?

    A. Ionization equilibrium constant, K_w **C.** Solubility product equilibrium constant, K_{sp}

    B. General equilibrium constant, K_{eq} **D.** Ionization equilibrium constant, K_i

                                        E. None of the above

58. Coal burning plants release the toxic gas, sulfur dioxide, into the atmosphere while nitrogen monoxide is released into the atmosphere via industrial processes and from combustion engines. Sulfur dioxide can also be produced in the atmosphere by the following equilibrium reaction:

$$SO_3 \ (g) + NO \ (g) + Heat \leftrightarrow SO_2 \ (g) + NO_2 \ (g)$$

Which of the following does NOT shift the equilibrium to the right?

    A. $[NO_2]$ decrease   **C.** Decrease the reaction chamber volume

    B. $[NO]$ increase   **D.** Temperature increase

                                        E. All of the above would shift the equilibrium to the right

---

Questions **59-62** are based on this graph and net reaction.

The reaction proceeds in two consecutive steps.

$$XY + Z \rightleftarrows XYZ \rightleftarrows X + YZ$$

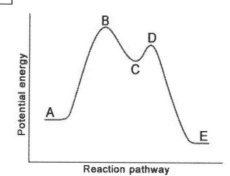

59. Where is the activated complex for this reaction?

    A. A   **B.** A and C   **C.** C   **D.** E   **E.** B and D

60. The activation energy for the reverse reaction is given by:

    A. A   **B.** B   **C.** C   **D.** D   **E.** D and C

61. The activation energy for the forward reaction is given by:

    A. A   **B.** B   **C.** C   **D.** D   **E.** B and D

62. The change in energy (ΔE) of the overall reaction is given by the difference between:

    A. A and B   **B.** A and E   **C.** A and C   **D.** B and D   **E.** B and C

63. Which statement is NOT correct for $aA + bB \rightarrow dD + eE$ whereby rate $= k[A]^q[B]^r$?

    A. The overall order of the reaction is q + r

    B. The exponents q and r are equal to the coefficients a and b, respectively

    C. The exponents q and r must be determined experimentally

    D. The exponents q and r are often integers

    E. The symbol k represents the rate constant

Copyright © 2015 Sterling Test Prep

64. If, at equilibrium, most of the reactants remain unreacted, the equilibrium constant has a numerical value that is:

A. slightly greater than 1.0

B. slightly less than 1.0

C. very large

D. equal to 0

E. very small

65. If the heat of reaction is exothermic, which of the following is always true?

A. Energy of the reactants is greater than the products

B. Energy of the reactants is less than the products

C. Reaction rate is fast

D. Reaction rate is slow

E. Energy of the reactants is equal to that of the products

66. Which factors decrease the rate of a reaction?

I. Lowering the temperature
II. Increasing the concentration of reactants
III. Adding a catalyst

A. I only   B. II only   C. III only   D. II and III only   E. I and III only

67. The rate constant for the conversion of methyl propanoate is $1 \times 10^{-4} \, s^{-1}$. A container of methyl propanoate gas has a partial pressure of 100 torr. After 6.4 hours (approx. 23,000 seconds), what is the partial pressure of methyl propanoate gas?

A. 1 torr   B. 10 torr   C. 5 torr   D. 30 torr   E. 12.5 torr

68. Which of the changes has no effect on the equilibrium for the reversible reaction?

$$PbI_2 \, (s) \leftrightarrow Pb^{2+} \, (aq) + 2I^- \, (aq)$$

A. Decreasing $[Pb^{2+}]$

B. Adding solid PbI_2

C. Decreasing $[I^-]$

D. Increasing $[Pb^{2+}]$

E. Adding solid $AgNO_3$

69. What does it mean when the position of an equilibrium is described as being "far to the left"?

A. The rate of the reverse reaction is greater than that of the forward reaction

B. Significant amounts of both products and reactants are present in the equilibrium mixture

C. Very few product molecules are present in the equilibrium mixture

D. Very few reactant molecules are present in the equilibrium mixture

E. Many product molecules are present in the equilibrium mixture

70. What is the activation energy?

A. Amount of energy required to separate reactants from the products

B. Energy difference between the reactants and the products

C. Amount of energy required to activate a phase change

D. Amount of energy released from the phase change

E. Minimum amount of energy to break the bonds in the reactants

Copyright © 2015 Sterling Test Prep

71. Which of the following increases the collision energy of gaseous molecules?

   A. Increasing the temperature   **C.** Increasing the concentration

   B. Adding a catalyst   **D.** All of the above   **E.** None of the above

72. For a reaction that has an equilibrium constant of 4.3×10^{-17} at 25 °C, the position at equilibrium is described as:

   A. equal amounts of reactants and products

   B. significant amounts of both reactants and products

   C. mostly products

   D. mostly reactants

   E. amount of product is slightly greater than reactants

73. If the temperature of the system is raised for the reaction $CO\ (g) + 2H_2O\ (g) \rightleftarrows CH_3OH\ (g)$ with $\Delta H < 0$, the equilibrium is:

   A. variable   **C.** shifted to the left

   B. unaffected   **D.** shifted to the right

                     E. shifted to the right by the square root of ΔH

74. Which of the changes shifts the equilibrium to the right for the following reversible reaction?

     $PbI_2\ (s) \leftrightarrow Pb^{2+}\ (aq) + 2\ I^-\ (aq)$

   A. add solid $Pb(NO_3)_2$ **C.** decrease $[I^-]$

   B. add solid PbI_2 **D.** increase $[Pb^{2+}]$ **E.** add solid $NaNO_3$

75. Which of the following is true regarding the K_{eq} expression?

   A. The value of K_{eq} is temperature dependent

   B. The K_{eq} expression contains only substances in the same physical state

   C. The K_{eq} expression was originally determined experimentally

   D. None of the above

   E. All of the above

76. Cyclobutane decomposes as $C_4H_8\ (g) \rightarrow 2C_2H_4\ (g)$. What is the rate of $C_2H_4\ (g)$ formation if during the course this reaction, the rate of consumption of C_4H_8 was 4.25×10^{-4} mol $L^{-1}\ s^{-1}$?

   A. 1.06×10^{-4} mol $L^{-1}\ s^{-1}$ **C.** 2.13×10^{-4} mol $L^{-1}\ s^{-1}$

   B. 8.50×10^{-4} mol $L^{-1}\ s^{-1}$ **D.** 4.25×10^{-4} mol $L^{-1}\ s^{-1}$

                     E. 1.81×10^{-3} mol $L^{-1}\ s^{-1}$

77. Which of the following changes most likely decreases the rate of a reaction?

   A. Increasing the reaction temperature

   B. Increasing the concentration of a reactant

   C. Increasing the activation energy for the reaction

   D. Decreasing the activation energy for the reaction

   E. Increasing the reaction pressure

Copyright © 2015 Sterling Test Prep

78. What is the term for a type of equilibrium in which all of the participating species are not in the same physical state?

    A. Homogeneous equilibrium   **C.** Concentration equilibrium

    B. Physical equilibrium   **D.** Heterogeneous equilibrium   **E.** None of the above

79. A rapid reaction is distinguished by:

    A. having a large activation energy

    B. having a small heat of reaction

    C. having a large heat of reaction

    D. being unaffected by catalysts

    E. having a small activation energy

80. What is the order of B in a reactant with the rate law of $k[A]^2$?

    A. 0   **C.** 2

    B. 1   **D.** ½   **E.** Cannot be determined from the information given

81. For the reaction, CO (g) + $2H_2O$ (g) $\rightleftarrows$ CH_3OH (g) with ΔH < 0, which factor increases the equilibrium yield for methanol?

    A. Decreasing the volume

    B. Decreasing the pressure

    C. Lowering the temperature of the system

    D. Decreasing the volume or lowering the temperature of the system

    E. None of the above

82. A catalyst functions by:

    A. increasing the rate of a reaction by increasing the heat of reaction

    B. increasing the rate of a reaction by increasing the activation energy of the reverse reaction only

    C. increasing the rate of a reaction by lowering the activation energy of the forward reaction only

    D. increasing the rate of a reaction by decreasing the heat of reaction

    E. increasing the rate of a reaction by providing an alternative pathway with a lower activation energy

X + Y → Z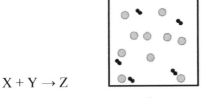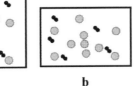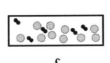

                      a   b   c

83. Which of the above reactions proceeds the slowest?

    A. a   **C.** c

    B. b   **D.** All proceed at same rate   **E.** Not enough information given

84. Which of the above reactions in the figures for question 83 proceeds the fastest?

    A. a   **C.** c

    B. b   **D.** All proceed at same rate   **E.** Not enough information given

 Copyright © 2015 Sterling Test Prep

85. Which of the following increases the collision frequency of molecules?

  A. Adding a catalyst

  B. Increasing the temperature

  C. Decreasing the concentration

  D. All of the above

  E. None of the above

86. For a chemical reaction at equilibrium, which always decreases the concentrations of the products?

  A. Decreasing the pressure

  B. Increasing the temperature and decreasing the temperature

  C. Increasing the temperature

  D. Decreasing the temperature

  E. Decreasing the concentration of a gaseous or aqueous reactant

87. Which statement is true?

  A. The equilibrium constant for a reaction can be determined in part from the activation energy

  B. The equilibrium concentrations of reactants can be determined in part from the activation energy

  C. The rate constant of a reaction can be determined in part from the activation energy

  D. The order of a reaction can be determined in part from the activation energy

  E. None of the above is true

88. From the data below, what is the order of the reaction with respect to reactant A?

Determining A Rate Law From Experimental Data

$A + B \rightarrow$ Products

Exp.	Initial [A]	Initial [B]	Initial Rate M/s
1	0.015	0.022	0.125
2	0.030	0.044	0.500
3	0.060	0.044	0.500
4	0.060	0.066	1.125
5	0.085	0.088	?

  A. Zero   **B.** First   **C.** Second   **D.** Third   **E.** Fourth

89. A chemical system is considered to have reached dynamic equilibrium when the:

  A. activation energy of the forward reaction equals the activation energy of the reverse reaction

  B. rate of production of each of the products equals the rate of their consumption by the reverse reaction

  C. frequency of collisions between the reactant molecules equals the frequency of collisions between the product molecules

  D. sum of the concentrations of each of the reactant species equals the sum of the concentrations of each of the product species

  E. rate of production of each of the product species equals the rate of consumption of each of the reactant species

90. $K_{eq} = 2$ with the initial concentrations of [A] = 4, [B] = 8 and [AB] = 16 at 25 °C.

$$AB + Heat \rightarrow A\ (g) + B\ (g)$$

After a stress has been absorbed by the reaction, the new equilibrium concentrations are: [A] = 2, [B] = 1, and [AB] = 64. The stress absorbed by the equilibrium system was:

  A. an increase in [AB]

  B. a decrease in [B]

  C. a decrease in [A]

  D. a change in the reaction temperature resulting in the change for K_{eq}

  E. a decrease in [A] and [B]

91. What is the equilibrium constant expression for the reaction $CO\ (g) + 2H_2\ (g) \leftrightarrow CH_3OH\ (l)$?

  A. $K_{eq} = 1 / [CO]\ [H_2]^2$

  B. $K_{eq} = [CO]\ [H_2]^2$

  C. $K_{eq} = [CH_3OH] / [CO]\ [H_2]^2$

  D. $K_{eq} = [CH_3OH] / [CO]\ [H_2]$

  E. None of the above

92. If a catalyst is added to the reaction, $CO + H_2O + heat \leftrightarrow CO_2 + H_2$, which direction does the equilibrium shift?

  A. To the left

  B. To the right

  C. No effect

  D. Not enough information

  E. Initially to the right but settles at ½ current equilibrium

93. With respect to A, what is the order of the reaction if the rate law = $k[A]^3[B]^6$?

  A. 2   **B.** 3   **C.** 4   **D.** 5   **E.** 6

94. If a reaction does not occur extensively and gives a low concentration of products at equilibrium, which of the following is true?

  A. The rate of the forward reaction is greater than the reverse reaction

  B. The rate of the reverse reaction is greater than the forward reaction

  C. The equilibrium constant is greater than one; that is, $K_{eq} \gg 1$

  D. The equilibrium constant is less than one; that is, $K_{eq} \ll 1$

  E. None of the above

95. Which of the following conditions characterizes a system in a state of chemical equilibrium?

  A. Rate of reverse reaction drops to zero

  B. Reactant molecules no longer react with each other

  C. Concentrations of reactants and products are equal

  D. Rate of forward reaction drops to zero

  E. Reactants are being consumed at the same rate they are being produced

Copyright © 2015 Sterling Test Prep

Questions **96** through **98** are based on the following Figure:

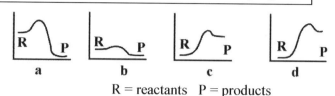

R = reactants   P = products

96. For the above energy profiles, which reaction has the highest activation energy?

   A. a   **B.** b   **C.** c   **D.** d   **E.** c and d

97. For the above energy profiles, which reaction has the lowest activation energy?

   A. a   **B.** b   **C.** c   **D.** d   **E.** a and b

98. For the above energy profiles, which reaction proceeds the slowest?

   A. a   **B.** b   **C.** c   **D.** d   **E.** c and d

99. What is the term for the difference in heat energy between the reactants and the products for a chemical reaction?

   A. Heat of reaction   **C.** Endothermic

   B. Exothermic   **D.** Activation energy   **E.** None of the above

100. If $K_{eq} = 6.1 \times 10^{-11}$, which statement is true?

   A. Both reactants and products are present   **C.** Mostly products are present

   B. The amount of reactants equals products   **D.** Cannot be determined

                                             E. Mostly reactants are present

Questions **101** and **102** are based on the following Equation:

$$H_2 + F_2 \rightleftharpoons 2HF$$

101. What happens when the concentration of H_2 is increased?

       I. equilibrium shifts to the left

       II. consumption of fluorine increases

       III. equilibrium shifts to the right

   A. I only   **B.** II only   **C.** III only   **D.** II and III only   **E.** I and III only

102. If a catalyst is added, which statement is true?

   A. The forward and backward reaction rates are increased by the same proportion

   B. The equilibrium is shifted to the more energetically favorable product

   C. The forward reaction is favored

   D. The equilibrium concentrations shift

   E. The reverse reaction is favored

103. Which change causes the greatest increase in the reaction rate if the rate law = $k[A][B]^2$?

A. Quadrupling [A]

B. Tripling [B]

C. Keeping the concentrations constant while decreasing the temperature

D. Doubling [B]

E. Doubling [A]

104. Increasing the temperature at which a chemical reaction occurs:

A. increases the reaction rate by lowering the activation energy

B. causes fewer reactant collisions to take place

C. lowers the activation energy, thus increasing the reaction rate

D. raises the activation energy, thus decreasing the reaction rate

E. increases the reaction rate by increasing reactant collisions per unit time

105. What is the equilibrium constant expression, K_i, for the following weak acid?

$$H_2S \, (aq) \leftrightarrow H^+ \, (aq) + HS^- \, (aq)$$

A. $K_i = [H^+]^2 \, [S^{2-}] \, / \, [H_2S]$

B. $K_i = [H_2S] \, / \, [H^+] \, [HS^-]$

C. $K_i = [H^+] \, [HS^-] \, / \, [H_2S]$

D. $K_i = [H^+]^2 \, [HS^-] \, / \, [H_2S]$

E. $K_i = [H_2S] \, / \, [H^+]^2 \, [HS^-]$

106. For $2SO_2 \, (g) + O_2 \, (g) \leftrightarrow 2SO_3 \, (g) + heat$, increasing pressure causes the equilibrium to:

A. remain unchanged, but the reaction mixture gets warmer

B. remain unchanged, but the reaction mixture gets cooler

C. shift to the right towards products

D. shift to the left towards reactants

E. pressure has no effect on equilibrium

107. Which method could be used to determine the rate law for a reaction?

I. Measure the initial rate of the reaction at several reactant concentrations

II. Graph the concentration of the reactants as a function of time

III. Deduce the reaction mechanism

A. I only   **B.** III only   **C.** II only   **D.** I, II and III   **E.** II and III only

108. What is the equilibrium constant expression for $CaCO_3 \, (s) \leftrightarrow CaO \, (s) + CO_2 \, (g)$?

A. $K_{eq} = [CO_2]$

B. $K_{eq} = 1 \, / \, [CO_2]$

C. $K_{eq} = [CaO] \, [CO_2] \, / \, [CaCO_3]$

D. $K_{eq} = [CaO] \, [CO_2]$

E. None of the above

109. Predict which reaction occurs at a faster rate for a hypothetical reaction $X + Y \rightarrow W + Z$.

Reaction	Activation energy	Temperature
1	low	low
2	low	high
3	high	high
4	high	low

A. 1   **B.** 2   **C.** 3   **D.** 4   **E.** 1 and 4

Copyright © 2015 Sterling Test Prep

110. If the graphs are for the same reaction, which most likely has a catalyst?

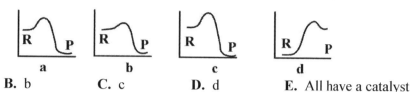

A. a   **B.** b   **C.** c   **D.** d   **E.** All have a catalyst

111. What is the general equilibrium constant expression, K_{eq}, for the reversible reaction:

$A + 3B \leftrightarrow 2C$?

A. $K_{eq} = [C]^2 / [A][B]^3$ **C.** $K_{eq} = [A][B]^3 / [C]^2$

B. $K_{eq} = [C] / [A][B]$ **D.** $K_{eq} = [A][B] / [C]$ **E.** None of the above

112. The following reaction is endothermic: $CaCO_3 (s) \leftrightarrow CaO (s) + CO_2 (g)$.

Which of the following causes the reaction to shift towards yielding more carbon dioxide gas?

  A. Increasing the pressure and decreasing the temperature of the system
  B. Increasing the pressure of the system
  C. Decreasing the temperature of the reaction
  D. Increasing the pressure and increasing the temperature of the system
  E. Increasing the temperature of the reaction

113. In chemical reactions, catalysts:

  A. shift the equilibrium
  B. become irreversibly consumed in the reaction
  C. lower the activation energy of the original reaction
  D. are depleted during the reaction
  E. increase the activation energy of the original reaction

114. What is the equilibrium constant expression, K_i, for the following weak acid?

$H_2CO_3 (aq) \leftrightarrow H^+ (aq) + HCO_3^- (aq)$

A. $K_i = [H^+]^2 [CO_3^{2-}] / [H_2CO_3]$ **C.** $K_i = [H^+][HCO_3^-] / [H_2CO_3]$

B. $K_i = [H_2CO_3] / [H^+][HCO_3^-]$ **D.** $K_i = [H^+]^2 [HCO_3^-] / [H_2CO_3]$

                                               E. $K_i = [H_2CO_3] / [H^+]^2 [CO_3^{2-}]$

115. Identify the reaction described by the following equilibrium expression:

$K = [H_2]^2 [O_2] / [H_2O]^2$

A. $2H_2O (g) \leftrightarrow 2H_2 (g) + O_2 (g)$ **C.** $H_2O (g) \leftrightarrow H_2 (g) + \frac{1}{2}O_2 (g)$

B. $H_2O (g) \leftrightarrow 2H (g) + O (g)$ **D.** $2H_2 (g) + O_2 (g) \leftrightarrow 2H_2O (g)$

                                             E. $2H_2O (g) \leftrightarrow H_2 (g) + O_2 (g)$

116. If the concentration of reactants decreases, which of the following is true?

  A. The amount of products increases   **C.** The rate of reaction decreases
  B. The heat of reaction decreases   **D.** All of the above
                                                E. None of the above

117. Which statement describes how a catalyst works?

A. increases ΔG **C.** decreases ΔH

B. increases E_{act} **D.** increases ΔH **E.** decreases E_{act}

118. What is the equilibrium law for CaO (s) + CO_2 (g) $\leftrightarrows$ $CaCO_3$ (s)?

A. $K_c = [CaCO_3] / [CaO]$ **C.** $K_c = [CaCO_3] / [CaO] [CO_2]$

B. $K_c = 1 / [CO_2]$ **D.** $K_c = [CO_2]$

        E. $K_c = [CaO] [CO_2] / [CaCO_3]$

119. The grams of products present after a chemical reaction reaches equilibrium:

A. must equal the grams of reactants present

B. may be less than, equal to, or greater than the grams of reactants present, depending upon the chemical reaction

C. must be greater than the grams of reactants present

D. must be less than the grams of reactants present

E. none of the above

120. Which of the following statements about catalysts is NOT true?

A. A catalyst does not change the energy of the reactants or the products

B. A catalyst can be used to speed up slow reactions

C. A catalyst can be consumed in a reaction as long as it is regenerated

D. A catalyst alters the rate of a chemical reaction

E. All of the above are true

121. Which of the following changes shifts the equilibrium to the left for the given reversible reaction?

$$SO_3 \, (g) + NO \, (g) + heat \leftrightarrow SO_2 \, (g) + NO_2 \, (g)$$

A. Decrease temperature   **C.** Increase [NO]

B. Decrease volume   **D.** Decrease [SO_2]   **E.** Add a catalyst

122. What is the expression for the equilibrium constant (K_{eq}) for the reaction is:

$$2NaIO_3 \, (s) \leftrightarrow 2NaI \, (s) + 3O_2 \, (g)$$

A. $K_{eq} = [NaIO_3]^2 / [NaI]^2[O_2]^3$ **C.** $K_{eq} = [O_2]^3 / [NaIO_3]^2$

B. $K_{eq} = [NaI]^2[O_2]^3 / [NaIO_3]^2$ **D.** $K_{eq} = [O_2]^3$ **E.** none of the above

123. If there is too much chlorine in the water, swimmers complain that their eyes burn. Consider the equilibrium found in swimming pools. Predict which increases the chlorine concentration.

$$Cl_2 \, (g) + H_2O \, (l) \leftrightarrow HClO \, (aq) \leftrightarrow H^+ \, (aq) + ClO^- \, (aq)$$

A. Decreasing the pH   **C.** Adding hypochlorous acid, $HClO$ (aq)

B. Adding hydrochloric acid, HCl (aq) **D.** Adding sodium hypochlorite, $NaClO$

        E. All of the above

       Copyright © 2015 Sterling Test Prep

124. The reaction, $2NO\ (g) + O_2\ (g) \rightarrow 2NO_2\ (g)$, was found to be first order in each of the two reactants and second order overall. The rate law is:

    A. rate $= k[NO]^2[O_2]^2$

    B. rate $= k[NO_2]^2[NO]^{-2}[O_2]^{-\frac{1}{2}}$

    C. rate $= k[NO][O_2]$

    D. rate $= k[NO]^2$

    E. rate $= k([NO][O_2])^2$

125. In writing an equilibrium constant expression, which of the following is NOT correct?

    A. Reactant concentrations are always placed in the denominator of the expression

    B. Concentrations of pure solids and pure liquids, when placed in the equilibrium expression, are never raised to any power

    C. Concentrations are always expressed as molarities

    D. Product concentrations are always placed in the numerator of the expression

    E. None of the above

126. Which of the following increases the amount of product formed from a reaction?

    A. Using a UV light catalyst

    B. Adding an acid catalyst

    C. Adding a metal catalyst

    D. None of the above

    E. All of the above

127. In the reaction $A + B \rightarrow AB$, which of the following does NOT increase the rate?

    A. increasing the temperature

    B. decreasing the temperature

    C. adding A

    D. adding B

    E. adding a catalyst

128. What is the equilibrium constant value if, at equilibrium, the concentrations are: $[NH_3] = 0.40$ M, $[H_2] = 0.12$ M and $[N_2] = 0.040$ M [at a certain temperature]?

$$2NH_3\ (g) \leftrightarrow N_2\ (g) + 3H_2\ (g)$$

    A. 6.3×10^{12} **B.** 7.1×10^{-7} **C.** 4.3×10^{-4} **D.** 3.9×10^{-3} **E.** 8.5×10^{-9}

129. Which of the changes listed below shifts the equilibrium to the left for the following reversible reaction: $PbI_2\ (s) \leftrightarrow Pb^{2+}\ (aq) + 2I^-\ (aq)$?

    A. Add $Pb(NO_3)_2\ (s)$

    B. Add NaCl (s)

    C. Decrease $[Pb^{2+}]$

    D. Decrease $[I^-]$

    E. Add nitric acid

130. Given the following reaction, the equilibrium expression will be:

$$4CuO\ (s) + CH_4\ (g) \leftrightarrow CO_2\ (g) + 4Cu\ (s) + 2H_2O\ (g)$$

    A. $[Cu]^4 / [CuO]^4$

    B. $[CH_4]^2 / [CO_2][H_2O]$

    C. $[CH_4] / [CO_2][H_2O]^2$

    D. $[CuO]^4 / [Cu]^4$

    E. $[CO_2][H_2O]^2 / [CH_4]$

131. Find the reaction rate for A + B → C:

Trial	$[A]_{t=0}$	$[B]_{t=0}$	Initial rate (M/s)
1	0.05 M	1.0 M	1×10^{-3}
2	0.05	4.0	16×10^{-3}
3	0.15	1.0	3×10^{-3}

A. rate = $k[A]^2[B]^2$

B. rate = $k[A][B]^2$

C. rate = $k[A]^2[B]$

D. rate = $k[A][B]$

E. rate = $k[A]^2[B]^3$

132. What is the general equilibrium constant expression, K_{eq}, for the reversible reaction?

$$2A + 3B \leftrightarrow C$$

A. $K_{eq} = [C] / [A]^2 [B]^3$

B. $K_{eq} = [C] / [A] [B]$

C. $K_{eq} = [A] [B] / [C]$

D. $K_{eq} = [A]^2 [B]^3 / [C]$

E. none of the above

133. What is the effect on the equilibrium after adding H_2O to the equilibrium mixture if CO_2 and H_2 react until equilibrium is established: $CO_2 (g) + H_2 (g) \leftrightarrow H_2O (g) + CO (g)$?

A. $[H_2]$ decreases and $[H_2O]$ increases

B. [CO] and $[CO_2]$ increase

C. $[H_2]$ decreases and $[CO_2]$ increases

D. Equilibrium shifts to the left

E. Equilibrium shifts to the right

> Questions **134** through **136** refer to the rate data
> for the conversion of reactants W, X, and Y to product Z.

Trial Number	Concentration (moles/L)			Rate of Formation of Z (moles/l·s)
	W	X	Y	
1	0.01	0.05	0.04	0.04
2	0.015	0.07	0.06	0.08
3	0.01	0.15	0.04	0.36
4	0.03	0.07	0.06	0.08
5	0.01	0.05	0.16	0.18

134. From the above data, what is the overall order of the reaction?

A. 3½   **B.** 4   **C.** 3   **D.** 2   **E.** 2½

135. From the above data, the order with respect to W suggests that the rate of formation of Z is:

A. dependent on [W]

B. independent of [W]

C. semidependent on [W]

D. unable to be determined

E. inversely proportional to [W]

136. From the above data, the magnitude of k for trial 1 is:

A. 20   **B.** 40   **C.** 60   **D.** 80   **E.** 90

Copyright © 2015 Sterling Test Prep

137. For the reaction, $2XO + O_2 \rightarrow 2XO_2$, data obtained from measurement of the initial rate of reaction at varying concentrations are:

Experiment	[XO]	$[O_2]$	Rate (mmol L^{-1} s^{-1})
1	0.010	0.010	2.5
2	0.010	0.020	5.0
3	0.030	0.020	45.0

The rate law is therefore:

A. rate = $k[XO][O_2]$

B. rate = $k[XO]^2 [O_2]^2$

C. rate = $k[XO]^2 [O_2]$

D. rate = $k[XO][O_2]^2$

E. rate = $k[XO]^2 / [O_2]^2$

138. 0.56 mol of NO and 0.38 mol of Br_2 are placed in a container and allowed to react until equilibrium, whereby 0.47 mol of NOBr are present.

$$2NO\ (g) + Br_2\ (g) \rightarrow 2NOBr\ (g)$$

What is the composition of the equilibrium mixture in terms of moles of each substance present?

A. 0.56 mol NO and 0.38 mol Br_2

B. 0.47 mol NO and 0.47 mol Br_2

C. 0.56 mol NO and 0.47 mol Br_2

D. 0.33 mol NO and 0.09 mol Br_2

E. 0.09 mol NO and 0.15 mol Br_2

139. What is the equilibrium constant expression, K_i, for the following weak acid?

$$H_2SO_3\ (aq) \leftrightarrow H^+\ (aq) + HSO_3^-\ (aq)$$

A. $K_i = [H_2SO_3] / [H^+] [HSO_3^-]$

B. $K_i = [H^+]^2 [SO_3^{2-}] / [H_2SO_3]$

C. $K_i = [H^+]^2 [HSO_3^-] / [H_2SO_3]$

D. $K_i = [H^+] [HSO_3^-] / [H_2SO_3]$

E. $K_i = [H_2SO_3] / [H^+]^2 [SO_3^{2-}]$

140. When a reaction system is at equilibrium:

A. rate in the forward and reverse directions are equal

B. rate in the forward direction is at a maximum

C. amounts of reactants and products are equal

D. rate in the reverse direction is at a minimum

E. reaction is complete and static

141. Which of the following drives the exothermic reaction to the right?

$$CH_4\ (g) + 2O_2\ (g) \leftrightarrow CO_2\ (g) + 2H_2O\ (g)$$

A. Addition of CO_2

B. Removal of CH_4

C. Increase in temperature

D. Decrease in temperature

E. Addition of H_2O

142. Which of the following is the equilibrium expression?

$$CS_2\ (g) + 4H_2\ (g) \leftrightarrow CH_4\ (g) + 2H_2S\ (g)$$

A. $K_{eq} = [CS_2][H_2]^4 / [CH_4][H_2S]^2$

B. $K_{eq} = [CH_4][H_2S]^2 / [CS_2][H_2]$

C. $K_{eq} = [CH_4][H_2S]^2 / [CS_2][H_2]^4$

D. $K_{eq} = [CS_2][H_2]^4 / [CH_4][H_2S]^2$

E. $K_{eq} = [CS_2][H_2]^2 / [CH_4][H_2S]^4$

143. Which of the following is the equilibrium expression?

$$C\ (s) + 2H_2\ (g) \leftrightarrow CH_4\ (g)$$

A. $K_{eq} = [H_2]^2\ [C]\ /\ [CH_4]$

B. $K_{eq} = [CH_4]\ /\ [H_2]^2$

C. $K_{eq} = [C]\ [H_2]$

D. $K_{eq} = [CH_4]\ /\ [C]\ [H_2]^2$

E. $K_{eq} = [CH_4]\ /\ [H_2]$

144. Which of the following increases the collision frequency of molecules?

A. Increasing the concentration

B. Adding a catalyst

C. Decreasing the temperature

D. All of the above

E. None of the above

145. What effect does a catalyst have on an equilibrium?

A. It increases the rate of the forward reaction

B. It shifts the reaction to the right

C. It increases the rate at which equilibrium is reached without changing ΔG

D. It increases the rate at which equilibrium is reached and lowers ΔG

E. It slows the reverse reaction

146. If k is 4×10^{-4} for $Cl_2\ (g) \rightleftharpoons 2Cl\ (g)$ at 1000 K, and 1 M $Cl_2\ (g)$ is placed in a container, what is the concentration of $Cl\ (g)$ at equilibrium?

A. 0.01 M   **B.** 0.02 M   **C.** 0.04 M   **D.** 0.05 M   **E.** 0.06 M

147. What is the rate law for the reaction, $3B + C \rightarrow E + 2F$, whereby the data for three trials are?

Experiment	[B]	[C]	Rate (mol $L^{-1}\ s^{-1}$)
1	0.100	0.250	0.000250
2	0.200	0.250	0.000500
3	0.100	0.500	0.00100

A. rate = $k[B]^2[C]^2$

B. rate = $k[B]^2[C]$

C. rate = $k[B]^3[C]$

D. rate = $k[B][C]^2$

E. rate = $k[B][C]$

148. A mixture of 1.40 moles of A and 2.30 moles of B reacted. At equilibrium, 0.90 moles of A are present. How many moles of C are present at equilibrium?

$$3A\ (g) + 2B\ (g) \rightarrow 4C\ (g)$$

A. 2.7 moles   **B.** 1.3 moles   **C.** 0.09 moles   **D.** 1.8 moles   **E.** 0.67 moles

149. What is the equilibrium constant expression (K_{eq}) for the reversible reaction below?

$$2A \leftrightarrow B + 3C$$

A. $K_{eq} = [B]\ [C]^3\ /\ [A]^2$

B. $K_{eq} = [B]\ [C]\ /\ [A]$

C. $K_{eq} = [A]^2\ /\ [B]\ [C]^3$

D. $K_{eq} = [A]\ /\ [B]\ [C]$

E. none of the above

150. What is the equilibrium expression for the reaction $2CO\ (g) + O_2\ (g) \leftrightarrow 2CO_2\ (g)$?

A. $K_{eq} = 2[CO_2]\ /\ 2[CO][O_2]$

B. $K_{eq} = [CO][O_2]\ /\ [CO_2]$

C. $K_{eq} = [CO_2]^2\ /\ [CO]^2[O_2]$

D. $K_{eq} = [CO_2]\ /\ [CO] + [O_2]$

E. $K_{eq} = [CO]^2[O_2]\ /\ [CO_2]^2$

Copyright © 2015 Sterling Test Prep

151. What is the rate law when rates were measured at different concentrations for the dissociation of hydrogen gas: $H_2(g) \rightarrow 2H(g)$?

$[H_2]$	Rate M/s s^{-1}
1.0	1.4×10^5
1.5	2.8×10^5
2.0	5.6×10^5

  A. The rate law cannot be determined from the information given

  B. rate $= k[H]^2/[H_2]$ **D.** rate $= k[H_2]/[H]^2$

  C. rate $= k[H_2]^2$ **E.** rate $= k2[H]/[H_2]$

152. Which of the changes shifts the equilibrium to the left for the reversible reaction in an aqueous solution: $HNO_2(aq) \leftrightarrow H^+(aq) + NO_2^-(aq)$?

  A. Adding solid KNO_2 **C.** Increasing $[HNO_2]$

  B. Adding solid KCl   **D.** Increasing pH

                            E. Adding solid KOH

153. What does a chemical equilibrium expression of a reaction depend on?

  A. mechanism   **C.** stoichiometry

  B. stoichiometry and mechanism   **D.** rate

                            E. rate and mechanism

154. What is the K_{sp} for slightly soluble copper (II) phosphate in an aqueous solution?

$$Cu_3(PO_4)_2(s) \rightleftharpoons 3Cu^{2+}(aq) + 2PO_4^{3-}(aq)$$

  A. $K_{sp} = [Cu^{2+}]^3[PO_4^{3-}]^2$ **C.** $K_{sp} = [Cu^{2+}]^3[PO_4^{3-}]$

  B. $K_{sp} = [Cu^{2+}][PO_4^{3-}]^2$ **D.** $K_{sp} = [Cu^{2+}][PO_4^{3-}]$

                            E. $K_{sp} = [Cu^{2+}]^2[PO_4^{3-}]^3$

155. Which of the following is true if the concentration of reactants increases?

  A. Amount of product decreases   **C.** Heat of reaction increases

  B. Rate of reaction increases   **D.** All of the above

                            E. Rate of reaction decreases

156. A reaction vessel contains NH_3, N_2, and H_2 at equilibrium with $[NH_3] = 0.1$ M, $[N_2] = 0.2$ M, and $[H_2] = 0.3$ M. For decomposition, what is K for NH_3 to N_2 and H_2?

  A. $K = (0.2)(0.3)^3/(0.1)^2$ **C.** $K = (0.1)^2/(0.2)(0.3)^3$

  B. $K = (0.1)/(0.2)^2(1.5)^3$ **D.** $K = (0.2)(0.3)^2/(0.1)$

                            E. $K = (0.2)(0.3)^3/(0.1)$

157. Given data the rate of a reaction as affected by the concentration of the reactants, the reaction:

Experiment	[A]	[B]	[C]	Rate (mol L^{-1} hr^{-1})
1	0.200	0.100	0.600	5.0
2	0.200	0.400	0.400	80.0
3	0.600	0.100	0.200	15.0
4	0.200	0.100	0.200	5.0
5	0.200	0.200	0.400	20.0

A. zero order with respect to A

B. order for A is minus one (rate proportional to 1 / [A])

C. first order with respect to A

D. second order with respect to A

E. order for A cannot be determined from data

158. Chemical equilibrium is reached in a system when:

A. complete conversion of reactants to products has occurred

B. product molecules begin reacting with each other

C. reactant concentrations steadily decrease

D. reactant concentrations steadily increase

E. product and reactant concentrations remain constant

159. Dinitrogen tetraoxide decomposes to produce nitrogen dioxide. Calculate the equilibrium constant given the equilibrium concentrations at 100 °C: $[N_2O_4] = 0.800$ and $[NO_2] = 0.400$.

$$N_2O_4\,(g) \leftrightarrow 2NO_2\,(g)$$

A. $K_{eq} = 0.725$ **C.** $K_{eq} = 0.200$

B. $K_{eq} = 2.50$ **D.** $K_{eq} = 0.500$ **E.** $K_{eq} = 0.750$

160. Which change to this reaction system causes the equilibrium to shift to the right?

$$N_2\,(g) + 3H_2\,(g) \leftrightarrow 2NH_3\,(g) + heat$$

A. Heating the system   **C.** addition of NH_3 (g)

B. Removal of H_2 (g)   **D.** Lowering the temperature

          E. Addition of a catalyst

161. Which is NOT a difference between a first order and second order elementary reactions?

A. When concentrations of reactants are doubled, the rate of a first order reaction doubles, while the rate of a second order elementary reaction quadruples

B. The rate of a first order reaction is greater than the rate of a second order reaction because collisions are not required

C. A first order reaction is unimolecular, while a second order reaction is bimolecular

D. The half life of a first order reaction is independent of the starting concentration of the reactant, while the half life of a second order reaction depends on the starting concentration of the reactant

E. The rate of a first order reaction is less than the rate of a second order reaction because collisions are required

Copyright © 2015 Sterling Test Prep

162. Which is the correct equilibrium expression for the reaction $2Ag\ (s) + Cl_2\ (g) \leftrightarrow 2AgCl\ (s)$?

A. $K_{eq} = [AgCl] / [Ag]^2[Cl_2]$

B. $K_{eq} = [2AgCl] / [2Ag][Cl_2]$

C. $K_{eq} = [AgCl]^2 / [Ag]^2[Cl_2]$

D. $K_{eq} = 1 / [Cl_2]$

E. $K_{eq} = [2AgCl]^2 / [2Ag]^2[Cl_2]$

163. How does a catalyst increase the rate of a reaction?

A. It increases the energy difference between the reactants and products

B. It has no affect on the rate of the reaction

C. It is neither created nor consumed in a reaction

D. It raises the activation energy of the reactants, which makes the reaction proceed faster

E. It lowers the activation energy

164. If the temperature of a reaction increases, which of the following is true?

A. Amount of product decreases

B. Rate of reaction increases

C. Heat of reaction increases

D. All of the above

E. None of the above

165. Given the equation $xA + yB \rightarrow z$, the rate expression reaction in terms of the rate of change of the concentration of c with respect to time is:

A. $k[A]^x[B]^{yz}$

B. $k[A]^y[B]^x$

C. $k[A]^x[B]^y$

D. unable to determine

E. $k[A]^x[B]^y / [z]$

166. Which one of the following changes the value of the equilibrium constant?

A. Varying the initial concentrations of reactants

B. Varying the initial concentrations of products

C. Changing temperature

D. Adding a catalyst at the onset of the reaction

E. Adding other substances that do not react with any of the species involved in the equilibrium

167. Which conditions increase the rate of a chemical reaction?

    I.   Lowering the temperature of the chemical reaction

    II. Significantly increasing the concentration of one of the reactants

    III. Adding a catalyst to a reaction

    IV. Increasing the pressure on a gaseous reaction system

A. II and IV only

B. II and III only

C. I and IV only

D. II, III and IV only

E. III and IV only

168. Which change shifts the equilibrium to the right for the reversible reaction in an aqueous solution?

$$HNO_2\,(aq) \leftrightarrow H^+\,(aq) + NO_2^-\,(aq)$$

A. Add solid NaOH

B. Decrease $[NO_2^-]$

C. Decrease $[H^+]$

D. Increase $[HNO_2]$

E. All of the above

169. The position of the equilibrium for a system where $K = 4.6 \times 10^{-15}$ can be described as being favored for [ ]; the concentration of products is relatively:

A. the left; large

B. the left; small

C. the right; large

D. the right; small

E. neither direction; large

170. In addition to molecular collision, what conditions are required for a reaction?

A. Sufficient energy of collision and proper spatial orientation of the molecules

B. Sufficient energy of collision only

C. Proper spatial orientation of the molecules only

D. Sufficient temperature and sufficient duration of molecular contact

E. Sufficient energy of collision and sufficient duration of molecular content

171. Which of the changes shifts the equilibrium to the right for the following reversible reaction?

$$CO\,(g) + H_2O\,(g) \leftrightarrow CO_2\,(g) + H_2\,(g) + heat$$

A. increasing volume

B. increasing temperature

C. increasing $[CO_2]$

D. adding a catalyst

E. increasing $[CO]$

172. Increasing the temperature of a chemical reaction increases the rate of reaction because:

A. both the collision frequency and collision energies of reactant molecules increases

B. the collision frequency of reactant molecules increases

C. the activation energy increases

D. the activation energy decreases

E. the stability of the products increases

173. Heat is often added to chemical reactions performed in the laboratory to:

A. compensate for the natural tendency of energy to disperse

B. increase the rate at which reactants collide

C. allow a greater number of reactants to overcome the barrier of the activation energy

D. increase the energy of the reactant molecules

E. all of the above

174. Which shifts the equilibrium to the left for the reversible reaction in an aqueous solution?

$$HC_2H_3O_2\,(aq) \rightleftarrows H^+\,(aq) + C_2H_3O_2^-\,(aq)$$

A. add solid KOH

B. add solid KNO_3

C. increase pH

D. increase $[HC_2H_3O_2]$

E. add solid $KC_2H_3O_2$

Copyright © 2015 Sterling Test Prep

175. As the temperature of a reaction increases, the reaction rate:

A. increases, but the rate constant remains constant

B. increases, along with the rate constant

C. remains constant, but the rate constant increases

D. remains constant, as does the rate constant

E. is not affected

176. Write the mass action expression for the reaction: $4Cr\ (s) + 3CCl_4\ (g) \leftrightarrows 4CrCl_3\ (g) + 4C\ (s)$

A. $K_c = [C][CrCl_3]\ /\ [Cr][CCl_4]$

B. $K_c = [C]^4[CrCl_3]^4\ /\ [Cr]^4\ [CCl_4]^3$

C. $K_c = [CrCl_3]^4\ /\ [CCl_4]^3$

D. $K_c = [CrCl_3]\ /\ [CCl_4]$

E. $K_c = [CrCl_3] + [CCl_4]$

177. Calculate the equilibrium constant for the reaction if, at equilibrium, a 3.25 L tank contains 0.343 mol O_2, 0.0212 mol SO_3 and 0.00419 mol SO_2: $2SO_3\ (g) \leftrightarrow 2SO_2\ (g) + O_2\ (g)$

A. 4.14×10^{-3}

B. 4.44×10^{-2}

C. 5.32×10^{-2}

D. 1.28×10^{-2}

E. 4.62×10^{-4}

178. Which of the following is true if a reaction occurs extensively and yields a high concentration of products at equilibrium?

A. The rate of the reverse reaction is greater than the forward reaction

B. The rate of the forward reaction is greater than the reverse reaction

C. The equilibrium constant is less than one; $K_{eq} \ll 1$

D. The equilibrium constant is greater than one; $K_{eq} \gg 1$

E. None of the above

179. Which of the following conditions characterizes a system in a state of chemical equilibrium?

A. Product concentrations are greater than reactant concentrations

B. Reactant molecules no longer react with each other

C. Concentrations of reactants and products are equal

D. Rate of forward reaction has dropped to zero

E. Reactants are being consumed at the same rate they are being produced

180. For the reaction $CO\ (g) + Cl_2\ (g) \rightarrow COCl_2\ (g)$, the equilibrium constant is 125. What is the equilibrium constant of the reaction $2CO\ (g) + 2Cl_2\ (g) \rightarrow 2COCl_2\ (g)$?

A. 2.50×10^3

B. 250

C. 1.56×10^3

D. 1.56×10^4

E. 2.50×10^4

181. Given the following decomposition reaction: $PCl_5\ (g) \leftrightarrow PCl_3\ (g) + Cl_2\ (g)$. What is the equilibrium constant for this reaction if 0.84 moles of PCl_5 are placed in a 1.0 L container and, at equilibrium, 0.72 moles of PCl_5 remains?

A. 0.14   B. 0.020   C. 0.67   D. 0.72   E. 1.14

Copyright © 2015 Sterling Test Prep

182. Which changes shift the equilibrium to the product side for the following reaction at equilibrium?

SO_2Cl_2 (g) $\leftrightarrow$ SO_2 (g) + Cl_2 (g)

I. Addition of SO_2Cl_2 III. Removal of SO_2Cl_2

II. Addition of SO_2 IV. Removal of Cl_2

A. I, II and III   **B.** II and III   **C.** I and II   **D.** III and IV   **E.** I and IV

183. Which influences the rate of a chemical reaction?

A. catalyst   **C.** concentration
B. temperature   **D.** all of the above
                                   E. none of the above

184. The rate-determining step of a chemical reaction involves:

A. the fastest step   **C.** the molecules with the smallest molecular mass
B. the slowest step   **D.** the molecules with the greatest molecular mass
                                   E. charged molecules transformed into neutral molecules

185. The system, H_2 (g) + X_2 (g) $\leftrightarrows$ 2HX (g) has a value of 24.4 for K_C. A catalyst was introduced into a reaction within a 3.00 liter reactor containing 0.150 moles of H_2, 0.150 moles of X_2, and 0.600 moles of HX. The reaction proceeds in which direction?

A. to the right, Q > K   **C.** to the right, Q < K
B. to the left, Q > K   **D.** to the left, Q < K
                                    E. not possible to predict in direction

186. For a hypothetical reaction, A + B → C, predict which reaction occurs at the slowest rate from the reaction conditions from the data.

Reaction	Activation energy	Temperature
1	103 kJ/mol	15 °C
2	46 kJ/mol	22 °C
3	103 kJ/mol	24 °C
4	46 kJ/mol	30 °C

A. 1   **B.** 2   **C.** 3   **D.** 4   **E.** not enough information is given

187. Which of the changes has no effect on the equilibrium for the reversible reaction in an aqueous solution $HC_2H_3O_2$ (aq) $\leftrightarrow$ H^+ (aq) + $C_2H_3O_2^-$ (aq)?

A. Adding solid $NaC_2H_3O_2$ **C.** Increasing [$HC_2H_3O_2$]
B. Adding solid $NaNO_3$ **D.** Increasing [H^+]
                                           E. Adding solid NaOH

          Copyright © 2015 Sterling Test Prep

188. The function of a catalyst in a reaction system is to:

A. decrease the amount of reactants consumed
B. decrease the amount of energy consumed by the reaction
C. decrease the amount of heat produced
D. increase the yield of product
E. increase the rate of the reaction

189. What is an explanation for the observation that the reaction stops before all reactants are converted to products in the following reaction?

$$NH_3\,(aq) + HC_2H_3O_2\,(aq) \rightarrow NH_4^+\,(aq) + C_2H_3O_2^-\,(aq)$$

A. The catalyst is depleted
B. The reverse rate increases, while the forward rate decreases until they are equal
C. As [products] increases, the acetic acid begins to dissociate, stopping the reaction
D. As [reactants] decreases, NH_3 and $HC_2H_3O_2$ molecules stop colliding
E. As [products] increases, NH_3 and $HC_2H_3O_2$ molecules stop colliding

190. Which statement is NOT true regarding an equilibrium constant for a particular reaction?

A. Does not change as product is removed
B. Does not change as additional quantity of a reactant is added
C. Changes when a catalyst is added
D. Changes as the temperature increases
E. All are true statements

191. Which of the following changes shift the equilibrium to the right?

$$N_2\,(g) + 3H_2\,(g) \leftrightarrow 2NH_3\,(g) + 92.94 \text{ kJ}$$

1. Increasing the temperature
2. Decreasing the temperature
3. Removing NH_3
4. Adding NH_3
5. Removing N_2
6. Adding N_2

A. 2, 4, 6   **B.** 1, 3, 6   **C.** 1, 4, 5   **D.** 2, 3, 6   **E.** 1, 3, 5

192. Which changes shift(s) the equilibrium to the right for the reversible reaction in an aqueous solution: $HC_2H_3O_2\,(aq) \leftrightarrow H^+\,(aq) + C_2H_3O_2^-\,(aq)$?

A. Decreasing $[C_2H_3O_2^-]$
B. Adding solid NaOH
C. Increasing $[HC_2H_3O_2]$
D. Decreasing $[H^+]$
E. All of the above

193. Which is the rate law when the following experimental data was observed for $A + B \rightarrow C$?

Trial	$[A]_{t=0}$	$[B]_{t=0}$	Initial rate (M/s)
1	1.0 M	1.0 M	6.0×10^{-5}
2	2.0	1.0	1.2×10^{-4}
3	2.0	2.0	2.4×10^{-4}
4	2.0	4.0	4.8×10^{-4}

A. rate $= k[B]^2$
B. rate $= k[A][B]$
C. rate $= k[A]$
D. rate $= k[B]$
E. rate $= k[A]^2$

Copyright © 2015 Sterling Test Prep

194. Which statement about chemical equilibrium is NOT true?

   A. At equilibrium, the forward reaction rate equals the reverse reaction rate

   B. The same equilibrium state can be attained starting either from the reactant or product side of the equation

   C. Chemical equilibrium can only be attained by starting with reagents from the reactant side of the equation

   D. At equilibrium, [reactant] and [product] are constant over time

   E. At equilibrium, [reactant] and [product] may be different

195. At a given temperature, K = 46.0 for the reaction $4HCl\ (g) + O_2\ (g) \leftrightarrow 2H_2O\ (g) + 2Cl_2\ (g)$. At equilibrium, [HCl] = 0.150, $[O_2]$ = 0.395 and $[H_2O]$ = 0.625. What is the concentration of Cl_2 at equilibrium?

   A. 0.153 M   **B.** 0.444 M   **C.** 1.14 M   **D.** 0.00547 M   **E.** none of the above

196. Carbonic acid in blood: $CO_2\ (g) + H_2O\ (l) \leftrightarrow H_2CO_3\ (aq) \rightleftarrows H^+\ (aq) + HCO_3^-\ (aq)$ at equilibrium . If a person hyperventilates, the rapid breathing expels carbon dioxide gas. Which of the following decreases when a person hyperventilates.

   A. $[HCO_3^-]$ **C.** $[H_2CO_3]$

   B. $[H^+]$ **D.** None of the above   **E.** All of the above

197. The equilibrium expression for the reaction $A + 2B \leftrightarrow 2C + D$ is:

   A. $[C]^2[D] / [A][B]^2$ **C.** $[A]2[B] / 2[C][D]$

   B. $[A]2[B]^2 / 2[C]^2[D]$ **D.** $2[C][D] / [A]2[B]$

                                   E. $[A][B]^2 / [C]^2[D]$

198. Consider the following reaction: $H_2\ (g) + I_2\ (g) \rightarrow 2HI\ (g)$

At 160 K, this reaction has an equilibrium constant of 35. If, at 160 K, the concentration of hydrogen gas is 0.4 M, iodine gas is 0.6 M, and hydrogen iodide gas is 3 M:

   A. system is at equilibrium   **C.** [hydrogen iodide] increases

   B. [iodine] increases   **D.** [hydrogen iodide] decreases

                                   E. [hydrogen] increases

199. Which statement is true if $K_{eq} = 2.2 \times 10^{12}$?

   A. Mostly products are present   **C.** Mostly reactants are present

   B. There is an equal amount of reactants and products   **D.** Mone of the above

                                         E. Need more information

200. When a system is at equilibrium, the:

   A. reaction rate of the forward reaction is small compared to the reverse

   B. amount of products and reactants is exactly equal

   C. reaction rate of the forward reaction is equal to the rate of the reverse

   D. reaction rate of the reverse reaction is small compared to the forward

   E. reaction rate of the reverse reaction is large compared to the forward

                                                  Copyright © 2015 Sterling Test Prep

201. Which of the following is true before a reaction reaches chemical equilibrium?

A. The rate of the forward reaction is increasing, and the rate of the reverse reaction is decreasing

B. The rate of the forward reaction is decreasing, and the rate of the reverse reaction is increasing

C. The rates of the forward and reverse reactions are increasing

D. The rates of the forward and reverse reactions are decreasing

E. None of the above

202. The following data were collected for the reaction $A + B + C \rightarrow D + E$:

Trial	$[A]_{t=0}$	$[B]_{t=0}$	$[C]_{t=0}$	Initial rate (M/s)
1	0.1 M	0.1 M	0.1 M	2×10^{-2}
2	0.1	0.1	0.2	4×10^{-2}
3	0.1	0.2	0.1	8×10^{-2}
4	0.2	0.1	0.1	8×10^{-2}

What is the rate law for this reaction?

A. rate $= k[A][B][C] / [D][E]$ C. rate $= k[A][B]^2[C]^2$

B. rate $= k[A]^2[B][C]^2$ D. rate $= k[A][B][C]$ E. rate $= k[A]^2[B]^2[C]$

203. Calculate a value for K_c for the reaction $NOCl\,(g) + \frac{1}{2}O_2\,(g) \leftrightarrows NO_2\,(g) + \frac{1}{2}Cl_2\,(g)$, using the data:

$2NO\,(g) + Cl_2\,(g) \leftrightarrows 2NOCl\,(g)$ $K_c = 3.20 \times 10^{-3}$
$2NO_2\,(g) \leftrightarrows 2NO\,(g) + O_2\,(g)$ $K_c = 15.5$

A. 4.49   B. 0.343   C. 4.32×10^{-4} D. 1.33×10^{-5} E. 18.4

204. According to Le Chatelier's principle, which changes shift to the left the equilibrium for:

$N_2\,(g) + 3H_2\,(g) \leftrightarrow 2NH_3\,(g) + heat$

A. Decreasing the temperature   C. Increasing $[N_2]$

B. Increasing $[H_2]$ D. Decreasing the pressure on the system

                 E. Increasing $[N_2]$ and $[H_2]$

205. Which of the following is true after a reaction reaches chemical equilibrium?

A. The amount of reactants and products are constant

B. The amount of reactants and products are equal

C. The amount of reactants is increasing

D. The amount of products is increasing

E. None of the above

206. Which factors would increase the rate of a chemical reaction?

     I. Increasing the temperature
     II. Removing products as they form
     III. Adding a catalyst

A. I and II   B. II and III   C. I, II and III   D. I only   E. II only

207. Which of the following concentrations of CH_2Cl_2 should be used in the rate law for Step 2 if CH_2Cl_2 is a product of the fast (first) step and a reactant of the slow (second) step?

    A. $[CH_2Cl_2]$ at equilibrium

    B. $[CH_2Cl_2]$ in Step 2 cannot be predicted because Step 1 is the fast step

    C. Zero moles per liter

    D. $[CH_2Cl_2]$ after Step 1 is completed

    E. None of the above

208. Which of the changes has no effect on the equilibrium for the following reversible reaction?

$$SrCO_3\ (s) \leftrightarrow Sr^{2+}\ (aq) + CO_3^{2-}\ (aq)$$

    A. Adding solid Na_2CO_3

    B. Adding solid $NaNO_3$

    C. Increasing $[CO_3^{2-}]$

    D. Increasing $[Sr^{2+}]$

    E. Adding solid $Sr(NO_3)_2$

209. In which of the following equilibrium systems will the equilibrium shift to the left when the pressure of the system is increased?

    A. $H_2\ (g) + Cl_2\ (g) \leftrightarrow 2HCl\ (g)$

    B. $2SO_2\ (g) + O_2\ (g) \leftrightarrow 2SO_3\ (g)$

    C. $4NH_3\ (g) + 5O_2\ (g) \leftrightarrow 4NO\ (g) + 6H_2O\ (g)$

    D. $N_2\ (g) + 3H_2\ (g) \leftrightarrow 2NH_3\ (g)$

    E. Increased pressure causes all to shift to the left

210. For the reaction, $CO\ (g) + 2H_2O\ (g) \rightleftarrows CH_3OH\ (g)$ with $\Delta H < 0$, which factor decreases the magnitude of the equilibrium constant K?

    A. Decreasing the temperature of this system

    B. Decreasing volume

    C. Decreasing the pressure of this system

    D. All of the above

    E. None of the above

211. What is the equilibrium constant for $PCl_5\ (g) + 2NO\ (g) \leftrightarrows PCl_3\ (g) + 2NOCl\ (g)$, given the two reactions shown with their equilibrium constants?

$$PCl_3\ (g) + Cl_2\ (g) \leftrightarrows PCl_5\ (g) \qquad\qquad K_1$$
$$2NO\ (g) + Cl_2\ (g) \leftrightarrows 2NOCl\ (g) \qquad\qquad K_2$$

    A. K_1/K_2 **B.** $(K_1K_2)^1$ **C.** K_1K_2 **D.** K_2/K_1 **E.** $K_2 - K_1$

212. Which of the following statements is true concerning the equilibrium system, whereby S combines with H_2 to form hydrogen sulfide, a toxic gas from the decay of organic material?

$$S\ (g) + H_2\ (g) \leftrightarrow H_2S\ (g) \qquad K_{eq} = 2.8 \times 10^{-21}$$

    A. Almost all the starting molecules are converted to product

    B. Decreasing $[H_2S]$ shifts the equilibrium to the left

    C. Decreasing $[H_2]$ shifts the equilibrium to the right

    D. Increasing the volume of the sealed reaction container shifts the equilibrium to the right

    E. Very little hydrogen sulfide gas is present in the equilibrium

Copyright © 2015 Sterling Test Prep

213. What is the K_{sp} for slightly soluble gold (III) chloride in an aqueous solution for the reaction shown?

$$AuCl_3\,(s) \leftrightarrow Au^{3+}\,(aq) + 3Cl^-\,(aq)$$

A. $K_{sp} = [Au^{3+}]^3\,[Cl^-] / [AuCl_3]$

B. $K_{sp} = [Au^{3+}]\,[Cl^-]^3$

C. $K_{sp} = [Au^{3+}]^3\,[Cl^-]$

D. $K_{sp} = [Au^{3+}]\,[Cl^-]$

E. $K_{sp} = [Au^{3+}]\,[Cl^-]^3 / [AuCl_3]$

214. What is the equilibrium expression for $4NH_3\,(g) + 5O_2\,(g) \rightarrow 4NO\,(g) + 6H_2O\,(g)$?

A. $K_{eq} = [H_2O]^6[NO]^4 / [NH_3]^4[O_2]^5$

B. $K_{eq} = [H_2O]^6[NO]^4 / [NH_3]^4$

C. $K_{eq} = [NO]^4 / [NH_3]^4[O_2]^5$

D. $K_{eq} = [6H_2O][4NO] / [4NH_3]\,[5O_2]$

E. Cannot be determined from the information given

215. Hydrogen gas reacts with iron (III) oxide to form iron metal (which produces steel), as shown in the reaction below. Which statement is NOT correct concerning the equilibrium system?

$$Fe_2O_3\,(s) + 3H_2\,(g) + heat \leftrightarrow 2Fe\,(s) + 3H_2O\,(g)$$

A. Continually removing water from the reaction chamber increases the yield of iron

B. Decreasing the volume of hydrogen gas reduces the yield of iron

C. Lowering the reaction temperature increases the concentration of hydrogen gas

D. Increasing the pressure on the reaction chamber increases the formation of products

E. Decreasing the pressure on the reaction chamber favors the formation of products

216. Which of the following factors increases the rate of a chemical reaction?

A. Decreasing the concentration

B. Adding a catalyst

C. Decreasing the temperature

D. All of the above

E. None of the above

217. A chemical system is considered to have reached equilibrium when the:

A. rate of production of each of the product species is equal to the rate of consumption of each of the reactant species by the reverse reaction

B. rate of consumption of each of the product species by the reverse reaction is equal to the rate of production of each of the reactant species by the reverse reaction

C. sum of the concentrations of each of the reactant species is equal to the sum of the concentrations of each of the product species

D. rate of production of each of the product species is equal to the rate of consumption of each of the product species by the reverse reaction

E. rate of production of each of the product species by the forward reaction is equal to the rate of production of each of the reactant species by the reverse reaction

218. Which of the reactions at equilibrium shift to the left when the pressure is decreased?

A. $N_2\,(g) + 3H_2\,(g) \leftrightarrow 2NH_3\,(g)$

B. $N_2O_4\,(g) \leftrightarrow 2NO_2\,(g)$

C. $2HCl\,(g) \leftrightarrow H_2\,(g) + Cl_2\,(g)$

D. $2SO_3\,(g) \leftrightarrow 2SO_2\,(g) + O_2\,(g)$

E. All respond to the same degree under a change of pressure

219. What is the equilibrium reaction, given the solubility product expression for slightly soluble iron (III) chromate is $K_{sp} = [Fe^{3+}]^2 [CrO_4^{2-}]^3$?

   A. $Fe_2(CrO_4)_3 (s) \leftrightarrow 3Fe^{3+} (aq) + 2CrO_4^{2-} (aq)$
   B. $Fe_3(CrO_4)_2 (s) \leftrightarrow 2Fe^{3+} (aq) + 3CrO_4^{2-} (aq)$
   C. $Fe_2(CrO_4)_3 (s) \leftrightarrow Fe^{3+} (aq) + CrO_4^{2-} (aq)$
   D. $Fe_2(CrO_4)_3 (s) \leftrightarrow 2Fe^{3+} (aq) + 3CrO_4^{2-} (aq)$
   E. $Fe_3(CrO_4)_2 (s) \leftrightarrow 3Fe^{3+} (aq) + 2CrO_4^{2-} (aq)$

220. Consider the reaction: $2H_2 + 2NO \rightarrow N_2 + 2H_2O$.

What rate law is most consistent, given the rates were measured at different concentrations?

[H₂]	[NO]	Rate/M s⁻¹
0.1	0.3	225
0.2	0.3	450
0.3	0.1	80
0.2	0.1	50

   A. rate = $k[H_2]^2[NO]$ **C.** rate = $k[H_2]^3[NO]$
   B. rate = $k[H_2]^2[NO]^3$ **D.** rate = $k[H_2]^3[NO]^2$ **E.** rate = $k[H_2][NO]^2$

Copyright © 2015 Sterling Test Prep

Chapter 8.   THERMOCHEMISTRY

1. What happens to the kinetic energy of a gas molecule when the gas is heated?
   A. Depends on the gas
   B. Kinetic energy increases
   C. Kinetic energy decreases
   D. Kinetic energy remains constant
   E. None of the above

2. For an isolated system, what can be exchanged between the system and its surroundings?
   A. Neither matter nor energy
   B. Temperature only
   C. Matter only
   D. Energy only
   E. Both matter and energy

3. When liquids and gases are compared, liquids have [ ] compressibility compared to gases and a [ ] density.
   A. smaller… smaller
   B. greater… greater
   C. greater… smaller
   D. smaller… greater
   E. same…same

4. Which of the following is a form of energy?
   A. chemical
   B. mechanical
   C. electrical
   D. heat
   E. all of the above

5. Determine ΔH for the reaction $CH_4 (g) + 2O_2 (g) \rightarrow CO_2 (g) + 2H_2O (l)$ given the following:

   $$CH_4 (g) + 2O_2 (g) \rightarrow CO_2 (g) + 2H_2O (g) \quad \Delta H = -802 \text{ kJ/mol}$$

   $$2H_2O (g) \rightarrow 2H_2O (l) \qquad\qquad \Delta H = -88 \text{ kJ/mol}$$

   A. –890 kJ/mol
   B. –714 kJ/mol
   C. 714 kJ/mol
   D. 890 kJ/mol
   E. –914 kJ/mol

6. For a closed system, what can be exchanged between the system and its surroundings?
   A. Both matter and energy
   B. Neither matter nor energy
   C. Energy only
   D. Matter only
   E. Temperature only

7. Snow forms in the clouds when water vapor freezes without ever passing through the liquid phase in a process known as:
   A. deposition
   B. freezing
   C. condensation
   D. sublimation
   E. melting

8. Which of the following is a form of energy?
   A. nuclear   B. light   C. heat   D. mechanical   E. all of the above

9. A process or reaction that takes in heat from the surroundings is said to be:
   A. exothermic
   B. isothermal
   C. conservative
   D. endothermic
   E. endergonic

Copyright © 2015 Sterling Test Prep

10. What kind of system is represented when an investigator compresses gas within a leak-proof system by pushing down on the inside of a piston?

    A. isolated

    B. closed

    C. open

    D. endergonic

    E. endothermic

11. Which of the following terms does NOT involve the solid state?

    A. freezing

    B. sublimation

    C. evaporation

    D. melting

    E. none of the above

12. A fuel cell contains hydrogen and oxygen gas that react explosively and the energy converts water to steam which drives a turbine to turn a generator that produces electricity. The fuel cell and the turbine represent which forms of energy?

    A. Electrical and mechanical energy

    B. Electrical and heat energy

    C. Chemical and mechanical energy

    D. Chemical and heat energy

    E. Nuclear and mechanical energy

13. Which conditions result in a negative ΔG for a reaction?

    A. ΔH is negative and ΔS is positive

    B. ΔH is negative and ΔS is negative

    C. ΔS is positive

    D. ΔH is negative

    E. ΔH is negative and ΔS is zero

14. A process or reaction that releases heat to the surroundings is:

    A. isothermal

    B. exothermic

    C. endothermic

    D. conservative

    E. exergonic

15. A 500 ml beaker of distilled water is placed under a bell jar, which is then covered by a layer of opaque insulation. After several days, some of the water evaporated. The contents of the bell jar are what kind of system?

    A. endothermic   **B.** exergonic   **C.** closed   **D.** open   **E.** isolated

16. Heat is a measure of:

    A. temperature

    B. internal thermal energy

    C. average kinetic energy

    D. potential energy

    E. none of the above

17. Which of the following is NOT an endothermic process?

    A. Condensation of water vapor

    B. Boiling soup

    C. Water evaporating

    D. Ice melting

    E. All are endothermic

    Copyright © 2015 Sterling Test Prep

18. A fuel cell contains hydrogen and oxygen gas that react explosively and the energy converts water to steam which drives a turbine to turn a generator that produces electricity. The fuel cell and the steam represent which forms of energy?

    A. electrical and heat energy   **C.** chemical and heat energy

    B. electrical and chemical energy   **D.** chemical and mechanical energy

                                            E. nuclear and mechanical energy

19. Which is a state function?

    A. ΔG **B.** ΔH **C.** ΔS **D.** all of the above   **E.** none of the above

20. A small bomb has exploded inside a sealed concrete bunker. What kind of system are the contents of the bunker, if the shielding of the bunker results in no heat or vibrations being detected by anyone leaning against the exterior wall?

    A. exergonic   **B.** entropic   **C.** open   **D.** closed   **E.** isolated

21. In which of the following pairs of physical changes are both changes exothermic?

    A. Melting and condensation   **C.** Sublimation and evaporation

    B. Freezing and condensation   **D.** Freezing and sublimation

                                            E. None of the above

22. A fuel cell contains hydrogen and oxygen gas that react explosively and the energy converts water to steam which drives a turbine to turn a generator that produces electricity. What are the initial and final forms of energy?

    A. Chemical and electrical energy   **C.** Chemical and mechanical energy

    B. Nuclear and electrical energy   **D.** Chemical and heat energy

                                            E. Nuclear and mechanical energy

23. Calculate the standard enthalpy change for the reaction: $CaCO_3\,(s) \rightarrow CaO\,(s) + CO_2\,(g)$ using the standard heats of formation shown below:

Compound	ΔH_f°
$CaCO_3\,(s)$	−1206.5 kJ/mol
$CaO\,(s)$	−635.5 kJ/mol
$CO_2\,(g)$	−393.5 kJ/mol

    A. −571 kJ/mol   **C.** 177.5 kJ/mol

    B. −242 kJ/mol   **D.** 570 kJ/mol   **E.** −1,028.5 kJ/mol

24. Which scientific principle explains the observation that the amount of heat transfer accompanying a change in one direction is numerically equal but opposite in sign to the amount of heat transfer in the opposite direction?

    A. Law of Conservation of Mass and Energy   **C.** Avogadro's Law

    B. Law of Definite Proportions   **D.** Law of Conservation of Mass

                                             E. Law of Conservation of Energy

25. Once an object enters a black hole, astronomers consider it to have left the universe which means the universe is:

    A. entropic   **B.** isolated   **C.** closed   **D.** open   **E.** exergonic

26. Which temperature is the hottest?

    I. 100 °C   II. 100 °F   III. 100 K

    A. I only   **C.** III only
    B. II only   **D.** I and II are equal   **E.** II and III equal

27. Which of the following quantities is needed to calculate the amount of heat energy released as water turns to ice at 0 °C?

    A. Heat of condensation for water and the mass
    B. Heat of vaporization for water and the mass
    C. Heat of fusion for water and the mass
    D. Heat of solidification for water and the mass
    E. Heat of fusion for water only

28. A fuel cell contains hydrogen and oxygen gas that react explosively and the energy converts water to steam which drives a turbine to turn a generator that produces electricity. What energy changes are involved in the process?

    A. Mechanical → electrical energy   **C.** Heat → mechanical energy
    B. Chemical → heat energy   **D.** None of the above
                                    E. All of the above

29. A person immersed into 8 °C H_2O would experience hypothermia much more rapidly than if exposed to air of 8 °C because:

    A. immersion in water prevents perspiration from warming the body
    B. H_2O exhibits hydrogen bonding, while air does not
    C. H_2O conducts heat more effectively than air
    D. H_2O conducts heat less effectively than air
    E. air exhibits hydrogen bonding while H_2O does not

30. How many joules of heat must be removed to lower the temperature of a 36.5 g Al bar from 84.1 °C to 56.8 °C? (specific heat of Al = 0.908 J/g °C).

    A. 905 J   **B.** 225 J   **C.** 572 J   **D.** 1063 J   **E.** 888 J

31. A nuclear power plant uses radioactive uranium to convert water to steam, which drives a turbine that turns a generator to produce electricity. The uranium and the steam represent which forms of energy?

    A. Electrical and heat energy   **C.** Chemical and mechanical energy
    B. Nuclear and heat energy   **D.** Chemical and heat energy
                                    E. Nuclear and mechanical energy

                                          Copyright © 2015 Sterling Test Prep

32. The mathematical equation that expresses the first law of thermodynamics is:

   A. $\Delta H = q + w$ **C.** $\Delta H = \Delta E + p\Delta V$

   B. $\Delta H = q + \Delta E$ **D.** $\Delta H = \Delta E - p\Delta V$ **E.** $\Delta E = q + w$

33. What is the purpose of the hollow walls within a closed hollow walled container that is effective at maintaining the temperature inside?

   A. Traps air trying to escape from the box, which minimizes convection

   B. Acts as an effective insulator, which minimizes convection

   C. Acts as an effective insulator, which minimizes conduction

   D. Provides an additional source of heat for the container

   E. Reactions occur within the walls that maintain the temperature within the container

34. The specific heat of substance A is one fourth that of substance B. The temperature of both a 28.0 g sample of substance A and a 14.0 g sample of substance B were raised 15 °C. Compared to the heat absorbed by substance B, the heat absorbed by substance A was:

   A. equal to   **B.** one-half   **C.** twice   **D.** four times   **E.** square root

35. The ΔG of formation for $N_2(g)$ at 25 °C is:

   A. more information is needed   **C.** negative

   B. positive   **D.** 1 kJ/mol   **E.** 0 kJ/mol

36. Which of the following is true for a reaction where products have more stable bonds and more orderly arrangement than the reactants?

   A. ΔH is negative and ΔS is positive   **C.** ΔH is negative and ΔS is zero

   B. ΔH is positive and ΔS is negative   **D.** ΔH and ΔS are positive

                                           E. ΔH and ΔS are negative

37. The energy change occurring in a chemical reaction at constant pressure is:

   A. ΔE **B.** ΔG **C.** ΔS **D.** ΔH **E.** ΔP

38. Which component of an insulated vessel design minimizes heat conduction?

   A. Reflective interior coating   **C.** Double-walled construction

   B. Heavy-duty plastic casing   **D.** Tight-fitting, screw-on lid

                                         E. Heavy-duty aluminum construction

39. Which of the following temperatures is NOT possible?

   A. –200 °C   **B.** 0 °C   **C.** 25 K   **D.** –200 K   **E.** 0 °F

40. How much heat energy in Joules is required to heat 16.0 g of copper from 23.0 °C to 66.1 °C? Specific heat of Cu = 0.382 J/g °C.

   A. 450 J   **B.** 109 J   **C.** 322 J   **D.** 812 J   **E.** 263 J

41. A nuclear power plant uses ^{235}U to convert water to steam that drives a turbine which turns a generator to produce electricity. What are the initial and final forms of energy?

- **A.** Heat energy and electrical energy
- **B.** Nuclear energy and electrical energy
- **C.** Chemical energy and mechanical energy
- **D.** Chemical energy and heat energy
- **E.** Nuclear energy and mechanical energy

42. All of the statements regarding the symbol ΔH are correct, EXCEPT:

- **A.** referred to as a change in entropy
- **B.** referred to as a change in enthalpy
- **C.** referred to as heat of reaction
- **D.** represents the difference between the energy used in breaking bonds and the energy released in forming bonds during the chemical reaction
- **E.** has a negative value for an exothermic reaction

43. Which component of an insulated vessel design minimizes convection?

- **A.** Heavy–duty aluminum construction
- **B.** Reflective interior coating
- **C.** Tight–fitting, screw–on lid
- **D.** Double–walled construction
- **E.** Heavy–duty plastic casing

44. A sample of aluminum absorbed 9.86 J of heat and the temperature increased from 23.2 °C to 30.5 °C. What is the mass of the aluminum? The specific heat of aluminum is 0.90 J/g °C.

- **A.** 8.1 g
- **B.** 5.6 g
- **C.** 6.8 g
- **D.** 1.5 g
- **E.** 11.8 g

45. Which is true of an atomic fission bomb according to the conservation of mass and energy law?

- **A.** The mass of the bomb and the fission products are identical
- **B.** A small amount of mass is converted into energy
- **C.** The energy of the bomb and the fission products are identical
- **D.** The mass of the fission bomb is greater than the mass of the products
- **E.** None of the above

46. What is the effect on the energy of the activated complex and on the rate of the reaction when a catalyst is added to a chemical reaction?

- **A.** Energy of the activated complex increases and the reaction rate decreases
- **B.** Energy of the activated complex decreases and the reaction rate increases
- **C.** Energy of the activated complex and the reaction rate increase
- **D.** Energy of the activated complex and the reaction rate decrease
- **E.** Energy of the activated complex remains the same, while the reaction rate decreases

47. Which component of an insulated vessel design minimizes heat radiation?

- **A.** Tight–fitting, screw–on lid
- **B.** Heavy–duty aluminum construction
- **C.** Double–walled construction
- **D.** Reflective interior coating
- **E.** Heavy–duty plastic casing

Copyright © 2015 Sterling Test Prep

48. Which of the following best describes temperature?

    A. Temperature is the measure of the average amount of kinetic energy in a substance

    B. Temperature is the measure of the heat of an object

    C. Temperature is the measure of the total amount of energy in a substance

    D. All of the above

    E. None of the above

49. What is the final temperature after 336 J of heat energy is removed from 25.0 g of H_2O at 19.6 °C? Specific heat of H_2O = 4.184 J/g °C.

    A. 30.4 °C   **B.** 23.2 °C   **C.** 26.4 °C   **D.** 28.7 °C   **E.** 16.4 °C

50. What term refers to a chemical reaction that absorbs heat energy?

    A. endothermic   **C.** exothermic

    B. isothermal   **D.** exergonic   **E.** all of the above

51. To simplify comparisons, the energy value of fuels is expressed in units of:

    A. kcal/g   **B.** kcal/L   **C.** kcal   **D.** kcal/mol   **E.** some other unit

52. When considering how to keep a cup of coffee warm, an office worker places a lid on the cup. This action reduces the heat loss from:

    A. radiation   **C.** conduction

    B. sublimation   **D.** convection   **E.** all of the above

53. What is the heat capacity of 84.0 g of H_2O? Specific heat of H_2O = 4.184 J/g °C.

    A. 485 J/°C   **B.** 513 J/°C   **C.** 92.3 J/°C   **D.** 355 J/°C   **E.** 351 J/°C

54. What term refers to a chemical reaction that releases heat energy?

    A. isothermal   **C.** endothermic

    B. exothermic   **D.** exergonic   **E.** all of the above

55. According to the kinetic theory, which is NOT true of ideal gases?

    A. For a sample of gas molecules, average kinetic energy is directly proportional to temperature

    B. There are no attractive or repulsive forces between gas molecules

    C. All collisions among gas molecules are perfectly elastic

    D. There is no transfer of kinetic energy during collisions between gas molecules

    E. None of the above

56. Which statement below is always true for a spontaneous chemical reaction?

    A. $\Delta S_{sys} - \Delta S_{surr} = 0$ **C.** $\Delta S_{sys} + \Delta S_{surr} < 0$

    B. $\Delta S_{sys} + \Delta S_{surr} > 0$ **D.** $\Delta S_{sys} + \Delta S_{surr} = 0$ **E.** $\Delta S_{sys} - \Delta S_{surr} < 0$

Copyright © 2015 Sterling Test Prep

57. Which compound is NOT correctly matched with the predominant intermolecular force associated with that compound in the liquid state?

Compound	Intermolecular force
A. CH_3OH	hydrogen bonding
B. HF	hydrogen bonding
C. Cl_2O	dipole-dipole interactions
D. HBr	van der Waals interactions
E. CH_4	van der Waals interactions

58. Which of the following best describes heat?

A. Heat can be measured with a thermometer

B. Objects at the same temperature contain the same quantity of heat

C. Heat is energy that moves from high to low temperature objects

D. Heat is a measure of the temperature of an object

E. Heat is a measure of the average amount of energy in an object

59. How many grams of Ag can be heated from 23 °C to 36 °C when 22 g of Au cools from 95.5 °C to 26.4 °C? Specific heat of Ag = 0.240 J/g °C; Specific heat of Au = 0.130 J/g °C.

A. 63 g   **B.** 1.5×10^3 g   **C.** 44 g   **D.** 22 g   **E.** 66 g

60. Which statement best describes the reaction: $HCl\ (aq) + KOH\ (aq) \rightarrow KCl\ (aq) + H_2O\ (l)$

A. HCl and potassium hydroxide solutions produce potassium chloride and H_2O

B. HCl and potassium hydroxide solutions produce potassium chloride solution and H_2O

C. Aqueous HCl and potassium hydroxide produce aqueous potassium chloride and H_2O

D. HCl and potassium hydroxide produce potassium chloride and H_2O

E. An acid plus a base produce H_2O and a salt

61. Which reaction tends to be the most stable?

A. Isothermic reaction   **C.** Endothermic reaction

B. Exothermic reaction   **D.** All three are equally stable   **E.** All three are unstable

62. Which of the following properties of a gas is/are a state functions?

    I. temperature

    II. heat

    III. work

A. I only   **B.** I and II only   **C.** II and III only   **D.** I, II and III   **E.** II only

63. Calculate the mass of gold that requires 468 J to heat the sample from 21.6 °C to 33.2 °C? Specific heat of gold = 0.130 J/g °C

A. 6.72 g   **B.** 483 g   **C.** 262 g   **D.** 63.4 g   **E.** 310 g

     Copyright © 2015 Sterling Test Prep

64. Which of the statements below best describes the following reaction?

$$HNO_3\ (aq) + LiOH\ (aq) \rightarrow LiNO_3\ (aq) + H_2O\ (l)$$

A. Nitric acid and lithium hydroxide solutions produce lithium nitrate solution and water

B. Nitric acid and lithium hydroxide solutions produce lithium nitrate and water

C. Nitric acid and lithium hydroxide produce lithium nitrate and water

D. Aqueous solutions of nitric acid and lithium hydroxide produce aqueous lithium nitrate and water

E. An acid plus a base produce H_2O and a salt

65. Based on the reaction shown, which statement is true?

$$N_2 + O_2 \rightarrow 2NO \qquad \Delta H = 43.2\ kcal$$

A. 43.2 kcal are consumed when 1.00 mole of O_2 reacts

B. 43.2 kcal are produced when 1.00 mole of NO is produced

C. 43.2 kcal are consumed when 1.00 g of N_2 reacts

D. 43.2 kcal are consumed when 1.00 g of O_2 reacts

E. 43.2 kcal are consumed when 1.00 g of NO is produced

66. Which statement must be true for all objects that radiate heat?

A. Any object that gets warmer must be experiencing conduction or convection

B. First Law of Thermodynamics does not apply to radiation

C. First Law of Thermodynamics applies to convection

D. All objects are gradually getting colder

E. No object can reach the temperature of absolute zero

67. A chicken cutlet provides 6.90×10^2 food Calories (Cal). A food Calorie is equivalent to 4180 J of heat energy. The heat energy provided by the chicken cutlet would be sufficient to heat 59.1 kg of water by how many °C? Specific heat of water = 4.18 J/g °C.

A. 5.32 °C   **B.** 21.8 °C   **C.** 5.91 °C   **D.** 11.7 °C   **E.** 4.18 °C

68. Which of the statements best describes the following reaction?

$$HC_2H_3O_2\ (aq) + NaOH\ (aq) \rightarrow NaC_2H_3O_2\ (aq) + H_2O\ (l)$$

A. Acetic acid and sodium hydroxide solutions produce sodium acetate and water

B. Aqueous solutions of acetic acid and sodium hydroxide produce aqueous sodium acetate and water

C. Acetic acid and sodium hydroxide solutions produce sodium acetate solution and water

D. Acetic acid and sodium hydroxide produce sodium acetate and water

E. An acid plus a base produce H_2O and a salt

69. What is the volume of three moles of O_2 at STP?

A. 11.2 L   **B.** 22.4 L   **C.** 67.2 L   **D.** 32.0 L   **E.** 5.51. L

70. Which violates the first law of thermodynamics?

    I. An isolated pendulum that swings indefinitely

    II. A battery that indefinitely maintains its charge

    III. A refrigerator converts all of the interior heat removed to provide electricity to the room in which it is located

  A. I only   **B.** II only   **C.** III only   **D.** I, II and III   **E.** II and III only

71. Which of the following elements occurs naturally as diatomic molecules?

  A. chlorine gas   **C.** fluorine gas

  B. iodine vapor   **D.** all of the above   **E.** none of the above

72. Determine the value of ΔE°_{rxn} for this reaction, whereby the standard enthalpy of reaction (ΔH°_{rxn}) for $C_2H_2(g) + 2H_2(g) \rightarrow C_2H_6(g)$ is -311.5 kJ mol^{-1}.

  A. -306.5 kJ mol^{-1} **C.** -316.0 kJ mol^{-1}

  B. -318.0 kJ mol^{-1} **D.** -364.6 kJ mol^{-1} **E.** $+466$ kJ mol^{-1}

73. A process that is unfavorable for enthalpy, but favorable with respect to entropy, could:

  A. occur at any temperature

  B. not occur regardless of temperature

  C. occur at high temperatures, but not at lower temperatures

  D. occur at low temperatures, but not at higher temperatures

  E. none of the above

74. The combustion of one mole of a gas produces 212 calories of energy. Express the energy released in this reaction in kilojoules:

  A. 0.393 kJ   **B.** 0.887 kJ   **C.** 789 kJ   **D.** 348 kJ   **E.** 3.88 kJ

75. Which is a general guideline for balancing an equation?

  A. Check each reactant and product to verify the coefficients

  B. Balance polyatomic ions as a single unit

  C. Begin balancing with the most complex formula

  D. Write correct formulas for reactants and products

  E. All of the above

76. What is the ratio of the diffusion rates of H_2 gas to O_2 gas?

  A. 4:1   **B.** 2:1   **C.** 1:3   **D.** 1:4   **E.** 1:2

77. Which of the following is an intensive property?

    I. volume   II. density   III. specific heat

  A. III only   **C.** II and III only

  B. I and II only   **D.** I, II and III   **E.** II only

        Copyright © 2015 Sterling Test Prep

78. In a chemical reaction, the bonds being formed are:

A. more energetic than the ones broken

B. less energetic than the ones broken

C. the same as the bonds broken

D. different from the ones broken

E. none of the above

79. Which statement concerning temperature change as a substance is heated is NOT correct?

A. As a liquid is heated, its temperature rises until its boiling point is reached

B. During the time a liquid is changing to the gaseous state, the temperature gradually increases until all the liquid is changed

C. As a solid is heated, its temperature rises until its melting point is reached

D. During the time a solid melts to a liquid the temperature remains constant

E. The temperature remains the same during the phase change

80. Which is NOT a general guideline for balancing an equation?

A. Balance polyatomic ions as a single unit

B. Check each reactant and product to verify the coefficients

C. Write correct formulas for reactants and products

D. Begin balancing with the most complex formula

E. Change a subscript in a formula if the equation is not balanced

81. Consider the reaction: $N_2 + O_2 \rightarrow 2NO$. $\Delta H = 43.2$ kcal

When 50.0 g of N_2 reacts, [   ] kcal will be [   ].

A. 43.2… produced

B. 77.1… consumed

C. 77.1… produced

D. 86.4… consumed

E. 86.4… produced

82. As an extensive property, if the change in a value for the decomposition of 110 grams of a substance is –45 kJ, what is the change in this value when 330 grams decompose?

A. –15 kJ   **B.** –45 kJ   **C.** –135 kJ   **D.** –560 kJ   **E.** Cannot be determined

83. Which would be the correct units for a heat of condensation value?

A. Cal / g   **B.** g / °C   **C.** Cal / °C   **D.** J / °C   **E.** K / °C

84. Which is NOT a general guideline for balancing an equation?

A. Balance ionic compounds as a single unit

B. Balance polyatomic ions as a single unit

C. Begin balancing with the most complex formula

D. Write correct formulas for reactants and products

E. Check each reactant and product to verify the coefficients

85. The thermodynamic systems that have high stability tend to demonstrate:

A. maximum ΔH and maximum ΔS

B. maximum ΔH and minimum ΔS

C. minimum ΔH and maximum ΔS

D. minimum ΔH and minimum ΔS

E. none of the above

Copyright © 2015 Sterling Test Prep

86. As an intensive property, if the change in μ for the decomposition of 140 grams of a substance is –60 kJ/mol, what is the change in μ from the decomposition of 420 grams?

A. –20 kJ   **B.** –70 kJ   **C.** –140 kJ   **D.** –80 kJ   **E.** –60 kJ

87. What is an endothermic reaction?

A. It is a reaction that requires heat as a reactant
B. It is a reaction where there is a net absorption of energy from a reaction
C. It is a reaction where the products have more energy than the reactants
D. All of the above
E. None of the above

88. What is the coefficient (n) of O_2 gas for the balanced equation?

$$n\text{P}\ (s) + n\text{O}_2\ (g) \rightarrow n\text{P}_2\text{O}_3\ (s)$$

A. 1   **B.** 2   **C.** 3   **D.** 5   **E.** none of the above

89. What is the partial pressure due to CO_2 of a mixture of gases at 700 torr that contains 40% CO_2, 40% O_2 and 10% H by pressure?

A. 760 torr   **B.** 280 torr   **C.** 420 torr   **D.** 560 torr   **E.** 700 torr

90. The standard enthalpy (ΔH°) at 25 °C for $NH_3\ (g) + HCl\ (g) \rightarrow NH_4Cl\ (s)$ is –175.9 kJ mol^{-1}. Determine ΔE°_{rxn}:

A. –164.8 kJ mol^{-1} **C.** –173.4 kJ mol^{-1}
B. +5134 kJ mol^{-1} **D.** –180.9 kJ mol^{-1} **E.** –170.9 kJ mol^{-1}

91. Which of the following represent forms of internal energy?

        I. bond energy   II. thermal energy   III. gravitational energy

A. II only   **C.** II and III only
B. I and II only   **D.** III only   **E.** I, II and III

92. Which of the following would have the same numerical magnitude?

A. Heats of sublimation and deposition   **C.** Heats of fusion and deposition
B. Heats of solidification and condensation   **D.** Heats of sublimation and condensation
                                              E. None of the above

93. What is the coefficient (n) of O_2 gas for the balanced equation: $n\text{P}\ (s) + n\text{O}_2\ (g) \rightarrow n\text{P}_2\text{O}_5\ (s)$?

A. 1   **B.** 2   **C.** 4   **D.** 5   **E.** none of the above

94. A flask contains a mixture of O_2, N_2 and CO_2. The pressure exerted by N_2 is 320 torr, and exerted by CO_2 is 240 torr. If the total pressure of the gases is 740 torr, what is the percent pressure of O_2?

A. 20%   **B.** 66%   **C.** 33%   **D.** 24%   **E.** 75%

                                    Copyright © 2015 Sterling Test Prep

95. What does internal energy for an ideal gas depend upon?

        I. pressure   II. temperature   III. volume

A. I only   **C.** I and III only

B. II only   **D.** I, II and III   **E.** III only

96. What is an exothermic reaction?

        I. It is a reaction where there is a net absorption of energy from a reaction

        II. It is a reaction where the products have more energy than the reactants

        III. It is a reaction that requires heat as a reactant

A. I and II only   **C.** II and III only

B. I and III only   **D.** I, II and III   **E.** none of the above

97. What is the coefficient (n) of P for the balanced equation: $nP\ (s) + nO_2\ (g) \rightarrow nP_2O_5\ (s)$?

A. 1   **B.** 2   **C.** 4   **D.** 5   **E.** none of the above

98. Calculate the enthalpy for the reaction $N_2 + O_2 \rightarrow 2NO$, given the bond dissociation energies of $N_2 = 226$ kcal/mol, $O_2 = 199$ kcal/mol and NO = 145 kcal/mol.

A. 135 kcal/mol   **C.** 280 kcal/mol

B. −135 kcal/mol   **D.** −280 kcal/mol   **E.** 199 kcal/mol

99. The internal energy of the system increases for which of the following situations?

        I. A clay ball is dropped and sticks to the ground

        II. Hydrogen and oxygen undergo an exothermic reaction to form H_2O

        III. A race car is driven at a constant speed while considering the effect of air resistance

A. II only   **C.** I and III only

B. II and III only   **D.** I, II and III   **E.** III only

100. Which of the following quantities are necessary in calculating the amount of heat energy required to change liquid H_2O at 75 °C to steam at 110 °C?

A. Specific heat of ice and specific heat of H_2O

B. Heat of fusion for H_2O and heat of condensation for H_2O

C. Specific heat of steam and heat of fusion for H_2O

D. Specific heat of H_2O, specific heat of steam, and heat of vaporization for H_2O

E. Specific heat of H_2O and specific heat of steam

101. What is the coefficient (n) of N_2 gas for the balanced equation: $nN_2\ (g) + nH_2\ (g) \rightarrow nNH_3\ (g)$?

A. 1   **B.** 2   **C.** 3   **D.** 4   **E.** None of the above

102. Which statement is true for a chemical reaction, whereby ΔH is < 0 and ΔS is < 0?

  A. Reaction may or may not be spontaneous, but spontaneity is favored by low temperatures
  B. Reaction may or may not be spontaneous, but spontaneity is favored by high temperatures
  C. Reaction must be spontaneous, regardless of temperature and becomes even more so at higher temperatures
  D. Reaction must be spontaneous, regardless of temperature and becomes even more so at lower temperatures
  E. No conclusion can be made from the limited information presented

103. Where does the energy released during an exothermic reaction originate from?

  A. Surroundings
  B. Kinetic energy of the reacting molecules
  C. Potential energy of the reacting molecules
  D. Thermal energy of the reactants
  E. Kinetic energy of the surrounding

104. If it takes energy to break bonds and you gain energy in the formation of bonds, how can some reactions be exothermic, while others are endothermic?

  A. Some products have more energy than others and they always require energy to be formed
  B. Some reactants have more energetic bonds than others and they will always release energy
  C. It is the total number of bonds that matters. Sometimes you create more bonds than you break and, since all bonds have same amount of energy, you gain or lose energy depending on the number of bonds
  D. It is the total amount of energy that matters. Sometimes some bonds are stronger than others and so you gain or lose energy when you form them
  E. None of the above

105. What is the coefficient (n) of H_2 gas for the balanced equation: $nN_2\ (g) + nH_2\ (g) \rightarrow nNH_3\ (g)$?

  A. 1   **B.** 2   **C.** 3   **D.** 4   **E.** none of the above

106. What is $\Delta H°$ for the decomposition of methane to C and diatomic gases?

  $CH_4\ (g) + 2O_2\ (g) \rightarrow CO_2\ (g) + 2H_2O\ (g)\ \Delta H = -191.8$ kJ/mole

  A. −361.4 kJ   **B.** 17.8 kJ   **C.** −13.6 kJ   **D.** 361.4 kJ   **E.** 33.4 kJ

107. Which expression yields the same value for T_1 and T_2 when both temperatures are in Celsius or Kelvin?

  A. T_1/T_2 **B.** T_1T_2 **C.** $T_1 - T_2$ **D.** $T_1 + T_2$ **E.** $T_1T_2/2$

108. How much energy is required to change 12.9 g of solid Cu to molten Cu at 1083 °C (melting point)? Heat of fusion for Cu = 205 J/g.

  A. 1320 J   **B.** 1870 J   **C.** 1430 J   **D.** 3540 J   **E.** 2640 J

Copyright © 2015 Sterling Test Prep

109. What is the coefficient (n) of NH_3 for the balanced equation nN_2 (g) + nH_2 (g) → nNH_3 (g)?

  A. 1   **B.** 2   **C.** 3   **D.** 4   **E.** None of the above

110. A quantity of an ideal gas occupies 300 cm^3 at 27 °C. Find its volume at −173 °C at constant pressure.

  A. 50 cm^3 **B.** 150 cm^3 **C.** 100 cm^3 **D.** 900 cm^3 **E.** 200 cm^3

111. How much heat is required to raise 5 grams of a material by 10 K if it requires 40 calories to raise 5 grams of this material by 10 °C?

  A. 0.0040 cal   **B.** 40 cal   **C.** 5 cal   **D.** 50 cal   **E.** 4.0 cal

112. From the energy profile graphs, which reaction is endothermic?

  R= reactants   P= products

  A. a   **B.** b   **C.** c   **D.** d   **E.** None of the above

113. What is the product for the combination reaction Li (s) + O_2 (g) → ?

  A. LiO_2 **B.** Li_2O_3 **C.** LiO   **D.** Li_3O_2 **E.** Li_2O

114. What is the term for a reaction when the bonds formed during the reaction are stronger than the bonds broken?

  A. endergonic   **C.** endothermic
  B. exergonic   **D.** exothermic   **E.** spontaneous

115. If a stationary gas has a kinetic energy of 500 J at 25 °C, what is its kinetic energy at 50 °C?

  A. 125 J   **B.** 450 J   **C.** 540 J   **D.** 1080 J   **E.** 270 J

116. How much heat must be absorbed to evaporate 14 g of NH_3 at −33 °C (condensation point)? Heat of condensation for NH_3 = 1380 J/g.

  A. 62,000 J   **C.** 91,000 J
  B. 46,000 J   **D.** 19,000 J   **E.** 1,380 J

117. What is the product for the combination reaction Ca (s) + O_2 (g) → ?

  A. Ca_2O_3 **B.** CaO_2 **C.** Ca_2O **D.** CaO   **E.** Ca_3O_2

118. An ideal gas fills a closed rigid container. As the number of moles of gas in the chamber is increased at a constant temperature:

  A. volume increases   **C.** pressure decreases
  B. pressure remains constant   **D.** the effect on pressure cannot be determined
                                         E. pressure increases

---

119. When the temperature of a suspension bridge changes from –10 °C on a winter night to 0 °C during the daylight, an iron bolt expands in diameter from 1.0000 cm to 1.0022 cm. What is the coefficient of linear expansion for iron? (Note: $\Delta L/L = a\Delta T$.)

A. $(2.2 \times 10^{-4})\,(15)$

B. $(2.2 \times 10^{-4})\,/\,15$

C. $15\,/\,(2.2 \times 10^{-4})$

D. $1\,/\,(15 \times 2.2 \times 10^{-4})$

E. $(2.2 \times 10^{-4})\,(15)\,/\,(2.2 \times 10^{-4})$

120. Given that the following energy profiles have the same scale, which of the reactions is the most exothermic?

R = reactants   P = products

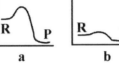

A. a   B. b   C. c   D. d   E. Cannot be determined

121. What is the product from heating magnesium metal and nitrogen gas?

A. Mg_2N B. Mg_2N_3 C. MgN D. MgN_2 E. Mg_3N_2

122. Consider the reaction: $C_3H_8 + 5O_2 \rightarrow 3CO_2 + 4H_2O + 488$ kcal

The reaction is [   ] and the sign of ΔH is [   ].

A. exothermic … negative

B. endothermic … negative

C. exothermic … positive

D. endothermic … positive

E. exothermic … neither positive nor negative

123. Which of the following expressions defines enthalpy? (q is heat; U is internal energy; P is pressure, V is volume)

A. $q - \Delta U$ B. $U + q$ C. q D. ΔU E. $U + PV$

124. What is the product from heating potassium metal and powdered phosphorus?

A. K_2P_3 B. K_3P C. KP_3 D. KP E. K_3P_2

125. Given the following data, what is the heat of formation for ethanol?

$C_2H_5OH + 3O_2 \rightarrow 2CO_2 + 3H_2O$
$\Delta H = 327.0$ kal/mole
$H_2O \rightarrow H_2 + \tfrac{1}{2}O_2$
$\Delta H = +68.3$ kcal/mole
$C + O_2 \rightarrow CO_2$
$\Delta H = -94.1$ kcal/mole

A. –66.1 kcal

B. –327.0 kcal

C. +66.1 kcal

D. +76.2 kcal

E. +327.0 kcal

126. For an ideal gas, ΔH depends upon:

I. density   II. temperature   III. volume

A. I only   B. II only   C. III only   D. I and II only   E. I and III only

Copyright © 2015 Sterling Test Prep

127. What is the heat of vaporization in J/g of an unknown liquid, if 6,823 J of heat is required to vaporize 58.0 g of it at its boiling point?

    A. 981 J/g   **B.** 28,100 J/g   **C.** 516 J/g   **D.** 118 J/g   **E.** 6,823 J/g

128. What is the product from heating cadmium metal and powdered sulfur?

    A. CdS_2 **B.** Cd_2S_3 **C.** CdS **D.** Cd_2S **E.** Cd_3S_2

129. Which of the following is true about Liquid A, if the vapor pressure of Liquid A is greater than that of Liquid B?

    A. Liquid A has a higher heat of fusion than B

    B. Liquid A has a higher heat of vaporization than B

    C. Liquid A forms stronger bonds than B

    D. Liquid A boils at a higher temperature than B

    E. Liquid A boils at a lower temperature than B

130. Under standard conditions, which reaction has the largest difference between the energy of reaction and enthalpy?

    A. C (*graphite*) → C (*diamond*)

    B. C (*graphite*) + O_2 (*g*) → C (*diamond*)

    C. 2C (*graphite*) + O_2 (*g*) → 2CO (*g*)

    D. C (*graphite*) + O_2 (*g*) → CO_2 (*g*)

    E. CO (*g*) + NO_2 (*g*) → CO_2 (*g*) + NO (*g*)

131. Given the energy profiles with the same scale, which of the reactions requires the most energy?

    R = reactants P = products

    A. a   **B.** b   **C.** c   **D.** d   **E.** None of the above

132. What are the predicted products for this decomposition reaction $LiHCO_3$ (*s*) → ?

    A. Li_2CO_3, H_2O and CO_2

    B. Li_2CO_3, H_2 and CO_2

    C. Li, H_2O and CO_2

    D. Li, H_2 and CO_2

    E. Li_2CO_3 and H_2O

133. Based on the reaction shown, which statement is true: $S + O_2 → SO_2 + 69.8$ kcal?

    A. 69.8 kcal are consumed when 32.1 g of sulfur reacts

    B. 69.8 kcal are produced when 32.1 g of sulfur reacts

    C. 69.8 kcal are consumed when 1 g of sulfur reacts

    D. 69.8 kcal are produced when 1 g of sulfur reacts

    E. 69.8 kcal are produced when 1 g of sulfur dioxide is produced

134. Consider the reaction P_4 (*s*) + $6Cl_2$ (*g*) → $4PCl_3$ (*l*). What is the term for this reaction, whereby $\Delta H° = -1287$ kJ/mol?

    A. nonspontaneous

    B. spontaneous

    C. exothermic

    D. endothermic

    E. endergonic

Copyright © 2015 Sterling Test Prep

135. Calculate the heat energy that must be removed from 17.6 grams of ammonia gas to condense it to liquid ammonia at its boiling point (−33 °C). The heat of vaporization for ammonia is 1380 J/g.

    A. 9,680 J   **B.** 0.0128 J   **C.** 78.4 J   **D.** 24,300 J   **E.** 723.4 J

136. What are the predicted products for this decomposition reaction $Zn(HCO_3)_2$ (s) → ?

    A. $ZnCO_3$, H_2 and CO_2 **C.** Zn, H_2 and CO_2

    B. $ZnCO_3$ and H_2O **D.** Zn, H_2O and CO_2 **E.** $ZnCO_3$, H_2O and CO_2

137. The equilibrium constant at 427 °C for the reaction N_2 (g) + $3H_2$ (g) ⇋ $2NH_3$ (g) is $K_p = 9.4 \times 10^{-5}$. Calculate the value of $\Delta G°$ for the reaction at this temperature:

    A. 56 kJ/mol   **B.** −56 kJ/mol   **C.** −33 kJ/mol   **D.** 33 kJ/mol   **E.** 1.3 J/mol

138. What is the standard enthalpy change for the reaction?

$$3P_4\,(s) + 18Cl_2\,(g) \rightarrow 12PCl_3\,(l)$$

$$P_4\,(s) + 6Cl_2\,(g) \rightarrow 4PCl_3\,(l) \qquad \Delta H° = -1289 \text{ kJ/mol}$$

    A. 426 kJ/mol   **C.** −366 kJ/mol

    B. −1345 kJ/mol   **D.** 1289 kJ/mol   **E.** −3837 kJ/mol

139. Which of the following reaction energies is the most endothermic?

    A. 540 kJ/mole   **C.** 125 kJ/mole

    B. −540 kJ/mole   **D.** −125 kJ/mole   **E.** Not enough information given

140. What are the predicted products for this decomposition reaction $Al(HCO_3)_3$ (s) → ?

    A. Al, H_2O and CO_2 **C.** $Al_2(CO_3)_3$, H_2O and CO_2

    B. $Al_2(CO_3)_3$, H_2 and CO_2 **D.** Al, H_2 and CO_2

    E. $Al_2(CO_3)_3$ and H_2O

141. If a substance is in the gas phase at STP, what occurs as the pressure of the surroundings is decreased at constant temperature?

    A. no phase change   **C.** evaporation

    B. sublimation   **D.** condensation   **E.** freezing

142. Given the data, what is the standard enthalpy change for $4PCl_3$ (l) → P_4 (s) + $6Cl_2$ (g)?

$$P_4\,(s) + 6Cl_2\,(g) \rightarrow 4PCl_3\,(l) \qquad \Delta H° = -1274 \text{ kJ/mol}$$

    A. 318.5 kJ/mol   **C.** −1917 kJ/mol

    B. −637 kJ/mol   **D.** 1274 kJ/mol   **E.** −2548 kJ/mol

143. How much heat is needed to convert 10.0 g of ice at −10 °C to H_2O (l) at 10 °C? Specific heat of ice = 2.09 J/(g °C); heat of fusion of ice = 334 J/g; specific heat of water = 4.18 J/g °C.

    A. 6,210 J   **B.** 3970 J   **C.** 2760 J   **D.** 1,070 J   **E.** 5230 J

 Copyright © 2015 Sterling Test Prep

144. What are the products for the single-replacement reaction Zn (s) + CuSO₄ (aq) → ?

    A. CuO and ZnSO₄   **C.** Cu and ZnSO₄

    B. CuO and ZnSO₃   **D.** Cu and ZnSO₃   **E.** no reaction

145. A reaction that is spontaneous can be described as:

    A. releasing heat to the surroundings

    B. proceeding without external influence once it has begun

    C. proceeding in both the forward and reverse directions

    D. having the same rate in both the forward and reverse directions

    E. increasing in disorder

146. Given that the $\Delta H°$ is 92.4 kJ/mol, what is the heat of formation of NH_3 (g) of the following reaction: $2NH_3$ (g) → N_2 (g) + $3H_2$ (g)?

    A. −92.4 kJ/mol   **C.** 46.2 kJ/mol

    B. −184.4 kJ/mol   **D.** 92.4 kJ/mol   **E.** −46.2 kJ/mol

147. Which of the following reaction energies is the least exothermic?

    A. 562 kJ/mole   **C.** 126 kJ/mole

    B. −562 kJ/mole   **D.** −126 kJ/mole   **E.** Not enough information given

148. Which solid compound is soluble in water?

    A. Na_2CO_3 **C.** $CuC_2H_3O_2$

    B. $AgNO_3$ **D.** None of the above   **E.** All of the above

149. If ΔG is negative, then the reaction is described as:

    A. spontaneous and endothermic   **C.** spontaneous with an increase in entropy

    B. spontaneous and exothermic   **D.** nonspontaneous because of the decrease in entropy

    E. spontaneous because of the decrease in entropy

150. The heat of formation of water vapor is:

    A. positive, but greater than the heat of formation for H_2O (*l*)

    B. positive and greater than the heat of formation for H_2O (*l*)

    C. negative, but greater than the heat of formation for H_2O (*l*)

    D. negative and less than the heat of formation for H_2O (*l*)

    E. positive and equal to the heat of formation for H_2O (*l*)

151. Which constant does C represent in the calculation used to determine how much heat is needed to convert 50 g of ice at −20 °C to steam at 300 °C?

(A)(50 g)(20 °C) + (heat of fusion)(50 g) + (B)(50 g)(100 °C) + (heat of vap.)(50 g) + (C)(50 g)(200 °C) = total heat.

    A. Heat capacity of water   **C.** Specific heat of water

    B. Specific heat of steam   **D.** Specific heat of ice   **E.** Heat capacity of steam

Copyright © 2015 Sterling Test Prep

152. Which solid compound is soluble in water?

   A. $AlPO_4$ **C.** $CaCO_3$

   B. $PbSO_4$ **D.** All of the above   **E.** None of the above

153. Which reaction is accompanied by an *increase* in entropy?

   A. $Na_2CO_3\,(s) + CO_2\,(g) + H_2O\,(g) \rightarrow 2NaHCO_3\,(s)$

   B. $BaO\,(s) + CO_2\,(g) \rightarrow BaCO_3\,(s)$

   C. $CH_4\,(g) + H_2O\,(g) \rightarrow CO\,(g) + 3H_2\,(g)$

   D. $ZnS\,(s) + 3/2\,O_2\,(g) \rightarrow ZnO\,(s) + SO_2\,(g)$

   E. $N_2\,(g) + 3H_2\,(g) \rightarrow 2NH_3\,(g)$

154. The greatest entropy is observed for which 10 g sample of CO_2?

   A. $CO_2\,(g)$ **C.** $CO_2\,(s)$

   B. $CO_2\,(aq)$ **D.** $CO_2\,(l)$ **E.** All are equivalent

155. A chemical reaction is when:

   A. two solids mix to form a heterogeneous mixture

   B. two liquids mix to form a homogeneous mixture

   C. one or more new compounds are formed by rearranging atoms

   D. a new element is formed by rearranging nucleons

   E. a liquid undergoes a phase change and produces a solid

156. Which solid compound is soluble in water?

   A. CuS **B.** Ag_3PO_4 **C.** $PbCrO_4$ **D.** $NiCO_3$ **E.** $Ba(OH)_2$

157. At equilibrium, increasing the temperature of the reaction likely:

   A. increases the heat of reaction

   B. decreases the heat of reaction

   C. increases the forward reaction

   D. decreases the forward reaction

   E. increases the heat of reaction and increases the forward reaction

158. Entropy can be defined as the amount of:

   A. equilibrium in a system

   B. chemical bonds that are changed during a reaction

   C. energy required to initiate a reaction

   D. energy required to rearrange chemical bonds

   E. disorder in a system

159. Which is ranked from lowest to highest entropy per gram of NaCl?

   A. $NaCl(s)$, $NaCl\,(l)$, $NaCl\,(aq)$, $NaCl\,(g)$

   B. $NaCl\,(s)$, $NaCl\,(l)$, $NaCl\,(g)$, $NaCl\,(aq)$

   C. $NaCl\,(g)$, $NaCl\,(aq)$, $NaCl\,(l)$, $NaCl\,(s)$

   D. $NaCl\,(s)$, $NaCl\,(aq)$, $NaCl\,(l)$, $NaCl\,(g)$

   E. $NaCl\,(l)$, $NaCl\,(s)$, $NaCl\,(g)$, $NaCl\,(aq)$

             Copyright © 2015 Sterling Test Prep

160. A chemical equation is a:

A. representation of the atoms undergoing a chemical equalization

B. chemical combination of equal numbers of reactants and products

C. sum of the masses of the products and reactants

D. shorthand notation for illustrating a chemical reaction

E. type of reaction that occurs only when energy is consumed by the reaction

161. Which constant does A represent in the calculation used to determine how much heat is needed to convert 50 g of ice at −20 °C to steam at 300 °C?

$(A)(50\text{ g})(20\text{ °C}) + (\text{heat of fusion})(50\text{ g}) + (4.18\text{ J/g °C})(50\text{ g})(100\text{ °C}) + (B)(50\text{ g}) + (C)(50\text{ g})(200\text{ °C}) = \text{total heat}.$

A. Specific heat of water

B. Heat of vaporization of water

C. Heat of condensation

D. Heat capacity of steam

E. Specific heat of ice

162. What phase change is observed by a researcher working with a sample of neon at 278 K and a pressure of 60 atm when the pressure is reduced to 38 atm?

A. liquid → solid

B. solid → liquid

C. solid → gas

D. liquid → gas

E. gas → liquid

163. At constant temperature and pressure, a negative ΔG indicates that the:

A. reaction is exothermic

B. reaction is fast

C. reaction is spontaneous

D. $\Delta S > 0$

E. reaction is endothermic

164. Which statement is true regarding entropy?

I. It is a state function

II. It is an extensive property

III. It has an absolute zero value

A. I only

B. III only

C. I and III only

D. I, II and III

E. I and II only

165. Is the potential energy of the reactants or the potential energy of the products higher in an endothermic reaction?

A. The potential energy of the reactants equals the potential energy of the products

B. Initially, the potential energy of the reactants is higher, but the potential energy of the products is higher in the later stages

C. The potential energy of the products is higher than the potential energy of the reactants

D. The potential energy of the reactants is higher than the potential energy of the products

E. Not enough information is provided to make any conclusions

166. Which of the following is a true statement regarding evaporation?

 A. Increasing the surface area of the liquid decreases the rate of evaporation

 B. The temperature of the liquid changes during evaporation

 C. Decreasing the surface area of the liquid increases the rate of evaporation

 D. Molecules with greater energy escape from the liquid

 E. Not enough information is provided to make any conclusions

167. What are the products for this double-replacement reaction $AgNO_3$ (*aq*) + $NaCl$ (*aq*) → ?

 A. $AgClO_3$ and $NaNO_2$ C. $AgCl$ and $NaNO_2$

 B. $AgCl$ and $NaNO_3$ D. Ag_3N and $NaClO_3$ E. $AgClO_3$ and $NaNO_3$

168. The noble gas Xe can exist in the liquid phase at 180 K, which is a temperature significantly greater than its normal boiling point, if:

 A. external pressure > vapor pressure of xenon

 B. external pressure < vapor pressure of xenon

 C. external pressure = partial pressure of water

 D. temperature is increased quickly

 E. temperature is increased slowly

169. What is the heat of reaction in the combustion of 60 grams of ethane from the data and balanced reaction below?

$$2C_2H_6 + 7O_2 \rightarrow 4CO_2 + 6H_2O \ (g)$$

$$\Delta H_f^o(C_2H_6) = -20.2 \text{ kcal/mol}$$
$$\Delta H_f^o(CO_2) = -94.1 \text{ kcal/mol}$$
$$\Delta H_f^o(H_2O \ (g)) = -57.8 \text{ kcal/mol}$$

 A. −682.8 kcal   C. −332.7 kcal

 B. −515.3 kcal   D. −243.2 kcal   E. −60.0 kcal

170. For the second law of thermodynamics, which statement(s) is/are true?

   I. Heat never flows from a cooler to a warmer object

   II. Heat cannot be converted completely to work in a cyclic process

   III. The entropy of the universe never decreases

 A. III only   C. II and III only

 B. I and II only   D. I, II and III   E. I and III only

171. Use the bond energies to determine if the reaction is exothermic: $H_2 + Cl_2 \rightarrow 2HCl$

H–H (436 kJ/mol); Cl–Cl (243 kJ/mol); H–Cl (431 kJ/mol)

 A. Exothermic with less than 50 kJ of energy released

 B. Endothermic with less than 50 kJ of energy absorbed

 C. Exothermic with more than 50 kJ of energy released

 D. Endothermic with more than 50 kJ of energy absorbed

 E. Endothermic equal to 50 kJ of energy absorbed

        Copyright © 2015 Sterling Test Prep

172. A nonvolatile liquid would have:

A. a highly explosive propensity

B. strong attractive forces between molecules

C. weak attractive forces between molecules

D. a high vapor pressure at room temperature

E. weak attractive forces within molecules

173. What are gases A and B likely to be if, at STP, a mixture of gas A and B has the average velocity of gas A twice that of B?

A. Ar and Kr   **B.** C and Ti   **C.** He and H   **D.** Mg and K   **E.** B and Ne

174. In the reaction, $2H_2(g) + O_2(g) \rightarrow 2H_2O(g)$, entropy is:

A. increasing   **C.** inversely proportional

B. the same   **D.** unable to be determined   **E.** decreasing

175. Which statement for the reaction $2NO(g) + O_2(g) \rightarrow 2NO_2(g)$ is true when $\Delta H° = -113.1$ kJ/mol and $\Delta S° = -145.3$ J/K mol?

A. Reaction is at equilibrium at 25 °C under standard conditions

B. Reaction is spontaneous at only high temperatures

C. Reaction is spontaneous at only low temperatures

D. Reaction is spontaneous at all temperatures

E. $\Delta G°$ becomes more favorable as temperature increases

176. When the system undergoes a spontaneous reaction, is it possible for entropy of a system to decrease?

A. No, because this violates the second law of thermodynamics

B. No, because this violates the first law of thermodynamics

C. Yes, but only if the reaction is endothermic

D. Yes, but only if the entropy gain of the environment is greater in magnitude than the magnitude of the entropy loss in the system

E. Yes, but only if the entropy gain of the environment is smaller in magnitude than the magnitude of the entropy loss in the system

177. From the given bond energies, how many kJ of energy are released or absorbed from the reaction of one mole of N_2 with three moles of H_2 to form two moles of NH_3?

$N{\equiv}N + H{-}H + H{-}H + H{-}H \rightarrow NH_3 + NH_3$

H–N (389 kJ/mol); H–H (436 kJ/mol); N≡N (946 kJ/mol);

A. –80 kJ/mol released   **C.** –946 kJ/mol released

B. +89.5 kJ/mol absorbed   **D.** +895 kJ/mol absorbed   **E.** +946 kJ/mol absorbed

178. The vapor pressure of $SnCl_4$ reaches 400 mmHg at 92 °C, the vapor pressure of SnI_4 reaches 400 mmHg at 315 °C, the vapor pressure of PBr_3 reaches 400 mmHg at 150 °C, and the vapor pressure of PCl_3 reaches 400 mmHg at 57 °C. At 175 °C, which substance would have the lowest vapor pressure?

A. PCl_3 **B.** $SnCl_4$ **C.** SnI_4 **D.** PBr_3 **E.** Hg

Copyright © 2015 Sterling Test Prep

179. If a chemical reaction is spontaneous, which must be negative?

   A. C_p **B.** ΔS **C.** ΔG **D.** ΔH **E.** K_{eq}

180. Consider the reaction $2CO\ (g) + O_2\ (g) \rightarrow CO_2\ (g) + 135.2$ kcal. This reaction is [   ] because the sign of ΔH is [   ].

   A. endothermic; negative   **C.** endothermic; positive

   B. exothermic; negative   **D.** exothermic; positive

                            E. exothermic; neither positive nor negative

181. When which quantity increases, does entropy increase?

      I. volume   II. moles   III. temperature

   A. II only   **C.** II and III only

   B. III only   **D.** I, II and III   **E.** I and III only

182. Decreasing the temperature of a liquid by 20 °C, at constant pressure, has what effect on the magnitude of its vapor pressure?

   A. no change   **C.** increase

   B. insufficient information given   **D.** inversely proportional   **E.** decrease

183. Which of the following conditions favors the formation of NO (g) in a closed container for the equilibrium: $N_2\ (g) + O_2\ (g) \leftrightarrow 2\ NO\ (g)$ $\Delta H = +181$ kJ/mol

   A. Increasing the temperature

   B. Decreasing the temperature

   C. Increasing the pressure

   D. Decreasing the pressure

   E. Decreasing the temperature and increasing the pressure

184. Which statement(s) is/are correct for the entropy?

      I. Higher for a sample of gas than for the same sample as liquid

      II. A measure of the *randomness* in a system

      III. Available energy for conversion into mechanical work

   A. I only   **B.** II only   **C.** III only   **D.** I and II only   **E.** I, II and III

185. Determine the equilibrium constant (K_c) at 25 °C for the reaction:

$$2NO\ (g) + O_2\ (g) \leftrightarrows 2NO_2\ (g)$$

$$(\Delta G°_{rxn} = -69.7 \text{ kJ/mol})$$

   A. 1.65×10^{12} **B.** 8.28×10^{-2} **C.** 2.60   **D.** 13.4   **E.** 6.07×10^{-13}

186. Which is NOT true for entropy in a closed system according to the equation $\Delta S = Q\ /\ T$?

   A. Entropy decreases as temperature decreases

   B. The equation is only valid for a reversible process

   C. Entropy changes due to heat transfer are greater at low temperatures

   D. Entropy of the system decreases as heat is transferred out of the system

   E. Entropy increases as temperature decreases

                                            Copyright © 2015 Sterling Test Prep

187. What role does entropy play in chemical reactions?

  A. The entropy change determines whether the reaction occurs spontaneously

  B. The entropy change determines whether the chemical reaction is favorable

  C. The entropy determines how much product is actually produced

  D. The entropy change determines whether the reaction is exothermic or endothermic

  E. The entropy determines how much reactant remains

188. The boiling point of a liquid is:

  A. the temperature where sublimation occurs

  B. the temperature where the vapor pressure of the liquid is less than the pressure over the liquid

  C. always 100 °C or greater

  D. the temperature where the rate of sublimation equals evaporation

  E. the temperature where the vapor pressure of the liquid equals the pressure over the liquid

189. What are the products for this double-replacement reaction $BaCl_2$ (*aq*) + K_2SO_4 (*aq*) → ?

  A. $BaSO_3$ and $KClO_4$ **C.** BaS and $KClO_4$

  B. $BaSO_4$ and KCl **D.** $BaSO_3$ and KCl **E.** $BaSO_4$ and $KClO_4$

190. Increased pressure results in:

  A. an increase in the forward reaction due to decreased volume

  B. an increase in the reverse reaction due to increased volume

  C. an increase in the forward reaction due to increased volume

  D. an increase in the reverse reaction due to decreased volume

  E. no effect since the reaction is at equilibrium

191. The process of H_2O (*g*) → H_2O (*l*) is nonspontaneous under pressure of 760 torr and temperatures of 378 K because:

  A. $\Delta H = T\Delta S$ **C.** $\Delta H > 0$

  B. $\Delta G < 0$ **D.** $\Delta H > T\Delta S$ **E.** $\Delta H < T\Delta S$

192. Using these bond energies, $\Delta H°$:

      C—C: 348 kJ   C=C: 612 kJ   C≡C: 960 kJ   C—H: 412 kJ

      C—O: 360 kJ   C=O: 743 kJ   H—H: 436 kJ   H—O: 463 kJ

Calculate the value of $\Delta H°$ of reaction for:

$$O=C=O\ (g) + 3\ H_2\ (g) \rightarrow CH_3\text{—}O\text{—}H\ (g) + H\text{—}O\text{—}H\ (g)$$

  A. –272 kJ   **B.** +272 kJ   **C.** –191 kJ   **D.** –5779 kJ   **E.** +5779 kJ

193. All of the statements regarding the symbol ΔG are true, EXCEPT it:

  A. identifies an endergonic reaction

  B. identifies an exothermic reaction

  C. predicts the spontaneity of a reaction

  D. refers to the free energy of the reaction

  E. describes the effect of both enthalpy and entropy on a reaction

Copyright © 2015 Sterling Test Prep

194. A negative ΔG signifies a spontaneous reaction when which conditions are constant?

    I. pressure   II. temperature   III. volume

    A. I only   **B.** II only   **C.** I and II only   **D.** I, II and III   **E.** III only

195. What happens to the entropy of a system as the components of the system are introduced to a larger number of possible arrangements, such as when liquid water transforms into water vapor?

    A. Entropy of a system is solely dependent upon the amount of material undergoing reaction

    B. Entropy of a system is independent of introducing the components of the system to a larger number of possible arrangements

    C. Entropy increases because there are more ways for the energy to disperse

    D. Entropy decreases because there are less ways in which the energy can disperse

    E. Entropy increases because there are less ways for the energy to disperse

196. Which statement about the boiling point of water is NOT correct?

    A. At sea level and at a pressure of 760 mmHg, the boiling point is 100 °C

    B. In a pressure cooker, shorter cooking times are required due to the change in boiling point

    C. The boiling point is greater than 100 °C in a pressure cooker

    D. The boiling point is less than 100 °C for locations at low elevations

    E. The boiling point is greater than 100 °C for locations at low elevations

197. If a chemical reaction has a positive ΔH and a negative ΔS, the reaction tends to be:

    A. at equilibrium   **C.** spontaneous

    B. nonspontaneous   **D.** unable to be determined   **E.** irreversible

198. The species in the reaction $KClO_3\,(s) \rightarrow KCl\,(s) + 3/2O_2\,(g)$ have the values for standard enthalpies of formation at 25 °C:

$$KClO_3\,(s),\ \Delta H^{\circ}_f = -391.2\ \text{kJ mol}^{-1}$$

$$KCl\,(s),\ \Delta H^{\circ}_f = -436.8\ \text{kJ mol}^{-1}$$

Make the assumption that, since the physical states do not change, the values of ΔH° and ΔS° are constant throughout a broad temperature range, and use this information to determine which of the following conditions may apply:

    A. The reaction is nonspontaneous at low temperatures, but spontaneous at high temperatures

    B. The reaction is spontaneous at low temperatures, but nonspontaneous at high temperatures

    C. The reaction is nonspontaneous at all temperatures over a broad temperature range

    D. The reaction is spontaneous at all temperatures over a broad temperature range

    E. It is not possible to make conclusion about spontaneity because of insufficient information

199. Which statement is NOT true for spontaneous reactions?

    A. Reaction rate is determined by the value of ΔG

    B. If the enthalpy change is unfavorable, the reaction occurs at a high temperature

    C. $\Delta G < 0$

    D. A catalyst does not affect ΔG

    E. Reaction is exergonic

 Copyright © 2015 Sterling Test Prep

200. Which condition(s) must be constant for $\Delta G = \Delta H - T\Delta S$ to be valid?

              I. pressure   II. temperature   III. volume

  A. I only   **B.** II only   **C.** III only   **D.** I, II and III   **E.** I and II only

201. The reaction rate is:

  A. the ratio of the masses of products and reactants

  B. the ratio of the molecular masses of the elements in a given compound

  C. the speed at which reactants are consumed or product is formed

  D. the balanced chemical formula that relates the number of product molecules to reactant molecules

  E. none of the above

202. Which substance would be expected to have the highest boiling point?

  A. Nonvolatile liquid

  B. Nonpolar liquid with van der Waal interactions

  C. Polar liquid with hydrogen bonding

  D. Nonpolar liquid

  E. Nonpolar liquid with dipole–induced dipole interactions

203. What are the products for this double-replacement reaction $AgNO_3\,(aq) + Li_3PO_4\,(aq) \rightarrow\ ?$

  A. Ag_3PO_4 and $LiNO_2$ **C.** Ag_3PO_3 and $LiNO_2$

  B. Ag_3PO_3 and $LiNO_3$ **D.** Ag_3P and $LiNO_3$

                                          E. Ag_3PO_4 and $LiNO_3$

204. The ΔG of a reaction is the maximum energy that the reaction releases to do:

  A. P–V work only   **C.** any type of work

  B. work and release heat   **D.** non P–V work only

                                            E. work and generate heat

205. If a chemical reaction has $\Delta H = X$, $\Delta S = Y$, $\Delta G = X - RY$ and occurs at R °K, the reaction is:

  A. spontaneous   **C.** nonspontaneous

  B. at equilibrium   **D.** irreversible   **E.** unable to be determined

206. Calculate the standard free energy change ($\Delta G°$) for $NO_2\,(g) + SO_2\,(g) \rightarrow NO\,(g) + SO_3\,(g)$ using the standard free energies of formation:

$$NO_2\,(g),\ \Delta G_f^o = +51.84 \text{ kJ mol}^{-1}$$

$$NO\,(g),\ \Delta G_f^o = +86.69 \text{ kJ mol}^{-1}$$

$$SO_2\,(g),\ \Delta G_f^o = -300.0 \text{ kJ mol}^{-1}$$

$$SO_3\,(g),\ \Delta G_f^o = -370.0 \text{ kJ mol}^{-1}$$

  A. –35.15 kJ   **B.** –104.9 kJ   **C.** –429.2 kJ   **D.** –619.6 kJ   **E.** –808.5 kJ

207. Consider the contribution of entropy to the spontaneity of the reaction

$$2Al_2O_3 (s) \rightarrow 4Al (s) + 3O_2 (g), \Delta G = +138 \text{ kcal}$$

As written, the reaction is [   ] and the entropy of the system [   ].

A. non-spontaneous… decreases

B. non-spontaneous… increases

C. spontaneous… decreases

D. spontaneous… increases

E. non-spontaneous… does not change

208. Which statement(s) is/are true for ΔG?

    I. ΔG of the universe is conserved

    II. ΔG of a system is conserved

    III. ΔG does not obey the first law of thermodynamics

A. I only   B. II only   C. III only   D. I and II only   E. I and III only

209. An Alka-Seltzer antacid tablet bubbles vigorously when placed in water, but only slowly when placed in an alcoholic beverage of the same temperature containing a 50:50 mix of alcohol and H_2O. Propose a probable explanation involving the relationship between the speed of a reaction and molecular collisions.

A. In a 50:50 mix there are fewer H_2O molecules for the antacid molecules to collide with

B. The tablet reacts chemically with H_2O, but not the alcohol

C. Alcohol molecules are more massive than H_2O molecules, hence they move slower and their collisions are not as forceful

D. The alcohol absorbs the carbon dioxide bubbles before they escape the liquid phase

E. In a 50:50 mix there are more H_2O molecules for the antacid molecules to collide with

210. Calculate the value of $\Delta H°$ of reaction for:

$$H_2C=CH_2 (g) + H_2 (g) \rightarrow CH_3—CH_3 (g)$$

The bond energies are:

    C—C: 348 kJ   C=C: 612 kJ   C≡C: 960 kJ

    C—H: 412 kJ   H—H: 436 kJ

A. –388 kJ   B. +224 kJ   C. –560 kJ   D. –124 kJ   E. –212 kJ

211. Which decreases as the strength of the attractive intermolecular force increases?

A. melting point

B. vapor pressure of a liquid

C. viscosity

D. normal boiling temperature

E. density

212. The bond dissociation energy is:

A. useful in estimating the enthalpy change in a reaction

B. the energy required to break a bond between two gaseous atoms

C. the energy released when a bond between two gaseous atoms is broken

D. A and B

E. all of the above

         Copyright © 2015 Sterling Test Prep

213. A spontaneous reaction:

   A. is endothermic and releases heat   **C.** occurs quickly

   B. cannot be endothermic   **D.** occurs slowly

                                   E. can do work on the surroundings

214. A catalyst changes which of the following?

   A. ΔS **B.** ΔG **C.** ΔH **D.** $E_{activation}$ **E.** $\Sigma(\Delta H_{products})$

215. When NaCl dissolves in water, what is the force of attraction between Na^+ and H_2O?

   A. ion–dipole   **C.** ion–ion

   B. hydrogen bonding   **D.** dipole–dipole   **E.** van der Waals

216. Identify the decreased ordering of attractions among particles in the three states of matter.

   A. solid < gas < liquid   **C.** solid > liquid > gas

   B. gas < solid < liquid   **D.** solid < liquid < gas   **E.** gas < liquid < solid

217. Which is true for the thermodynamic functions G, H and S in $\Delta G = \Delta H - T\Delta S$?

   A. G refers to the universe, H to the surroundings and S to the system

   B. G refers to the system, H to the system and S to the system

   C. G refers to the surroundings, H to the surroundings and S to the system

   D. G refers to the system, H to the system and S to the surroundings

   E. G refers to the system, H to the surroundings and S to the system

218. Which statement is the best explanation for the fact that the reaction rates of many thermodynamically spontaneous reactions are actually very slow?

   A. ΔS is negative   **C.** $\Delta G°$ for the reaction is positive

   B. K_P for the reaction < 1   **D.** Such reactions are endothermic

                                     E. The activation energy of the reaction is large

219. A solid sample at room temperature spontaneously sublimes forming a gas. This change in state is accompanied by which of the changes in the sample?

   A. Entropy decreases and energy increases   **C.** Entropy and energy decrease

   B. Entropy increases and energy decreases   **D.** Entropy and energy increase

                                            E. Entropy and energy are equal

220. What is true of a reaction if an endothermic reaction causes a decrease in ΔS of the system?

   A. Only occurs at low temperatures when ΔS is insignificant

   B. Occurs if coupled to a endergonic reaction

   C. Never occurs because it decreases ΔS of the system

   D. Never occurs because ΔG is positive

   E. None of the above

---

221. What can be deduced about the activation energy of a reaction that takes billions of years to go to completion and a reaction that takes only a fraction of a second.

  A. The slow reaction has high activation energy, while the fast reaction has low activation energy

  B. The slow reaction must have slow activation energy, while the fast reaction must have high activation energy

  C. The activation energy of both reactions is very low

  D. The activation energy of both reactions is very high

  E. The activation energy of both reactions is equal

222. Use the following bond dissociation energies to determine what the energy change is for: $2O_2 + CH_4 \rightarrow CO_2 + 2H_2O$?

  C—C   350 kJ/mole

  C—H   410 kJ/mole

  O=O   495 kJ/mole

  O—H   460 kJ/mole

  C=O   720 kJ/mole

  A. 880 kJ/mole   **C.** –650 kJ/mole

  B. –700 kJ/mole   **D.** –1158 kJ/mole   **E.** None of the above

223. Given the data, calculate ΔG°, for the reaction $H_2(g) + I_2(s) \rightarrow 2HI(g)$?

$$H_2(g),\ \Delta H_f^o = 0 \text{ kJ mol}^{-1},\ S^\circ = +130.6 \text{ J mol}^{-1}\text{K}^{-1}$$

$$I_2(s),\ \Delta H_f^o = 0 \text{ kJ mol}^{-1},\ S^\circ = +116.12 \text{ J mol}^{-1}\text{K}^{-1}$$

$$HI(g),\ \Delta H_f^o = +26 \text{ kJ mol}^{-1},\ S^\circ = +206 \text{ J mol}^{-1}\text{K}^{-1}$$

  A. +128.2 kJ   **B.** +2.7 kJ   **C.** –46.5 kJ   **D.** +64.1 kJ   **E.** –165.3 kJ

224. Since the physical states do not change, assume the values of ΔH and ΔS do not change as temperature is increased in the reaction below. Using the following values at 25 °C, calculate a value for the free energy change, ΔG_T^0.

$$CaCO_3(s) \rightarrow CaO(s) + CO_2(g) \text{ at 815 °C}$$

$$CaO(s),\ \Delta H_f^o = -635.5 \text{ kJ mol}^{-1},\ S^\circ = +40.0 \text{ J mol}^{-1}\text{K}^{-1}$$

$$CaCO_3(s),\ \Delta H_f^o = -1207 \text{ kJ mol}^{-1},\ S^\circ = +92.9 \text{ J mol}^{-1}\text{K}^{-1}$$

$$CO_2(g),\ \Delta H_f^o = -394 \text{ kJ mol}^{-1},\ S^\circ = +213.6 \text{ J mol}^{-1}\text{K}^{-1}$$

  A. +2.6 kJ   **B.** +42.5 kJ   **C.** +147.1 kJ   **D.** +312.2 kJ   **E.** +5.94 kJ

Copyright © 2015 Sterling Test Prep

ANSWER KEYS

Copyright © 2015 Sterling Test Prep

Chapter 1

1: D	26: E	51: B	76: C	101: B	126: A	151: B	176: A	201: B
2: E	27: D	52: E	77: D	102: C	127: C	152: E	177: C	202: E
3: D	28: E	53: B	78: B	103: B	128: D	153: D	178: E	203: D
4: B	29: B	54: A	79: D	104: B	129: D	154: C	179: B	204: D
5: A	30: B	55: E	80: E	105: D	130: E	155: C	180: B	205: A
6: A	31: C	56: B	81: C	106: E	131: C	156: B	181: D	206: A
7: E	32: A	57: A	82: D	107: A	132: B	157: C	182: A	207: B
8: B	33: B	58: D	83: B	108: D	133: D	158: A	183: E	208: E
9: C	34: E	59: B	84: A	109: E	134: E	159: D	184: C	209: C
10: C	35: D	60: E	85: C	110: D	135: D	160: E	185: B	210: C
11: E	36: B	61: D	86: B	111: C	136: B	161: A	186: E	211: D
12: B	37: A	62: C	87: E	112: B	137: B	162: A	187: A	212: C
13: A	38: C	63: D	88: D	113: A	138: E	163: D	188: C	213: A
14: E	39: C	64: E	89: A	114: C	139: A	164: E	189: A	214: D
15: D	40: D	65: A	90: C	115: E	140: B	165: A	190: D	215: E
16: C	41: E	66: C	91: E	116: D	141: D	166: C	191: E	216: C
17: B	42: A	67: D	92: A	117: C	142: E	167: E	192: D	217: B
18: A	43: A	68: C	93: E	118: E	143: C	168: D	193: B	218: B
19: A	44: C	69: E	94: B	119: D	144: A	169: A	194: C	219: A
20: C	45: E	70: A	95: D	120: A	145: C	170: B	195: A	220: A
21: D	46: D	71: A	96: A	121: B	146: E	171: B	196: C	221: E
22: A	47: A	72: D	97: C	122: A	147: A	172: D	197: D	222: B
23: D	48: C	73: C	98: B	123: A	148: A	173: E	198: E	223: B
24: E	49: B	74: B	99: E	124: C	149: C	174: B	199: B	224: D
25: C	50: D	75: C	100: A	125: E	150: B	175: C	200: D	225: A
								226: C

Copyright © 2015 Sterling Test Prep

Chapter 2

1: D	26: A	51: C	76: A	101: A	126: B	151: D
2: C	27: D	52: B	77: B	102: E	127: C	152: B
3: E	28: B	53: A	78: A	103: D	128: B	153: E
4: C	29: E	54: E	79: B	104: C	129: E	154: A
5: A	30: D	55: B	80: E	105: B	130: D	155: B
6: B	31: B	56: D	81: B	106: E	131: A	156: C
7: D	32: E	57: A	82: A	107: D	132: D	157: C
8: C	33: D	58: C	83: D	108: C	133: C	158: B
9: B	34: B	59: B	84: A	109: D	134: E	159: E
10: A	35: A	60: E	85: E	110: B	135: A	160: A
11: E	36: C	61: A	86: C	111: C	136: D	161: D
12: A	37: A	62: D	87: E	112: B	137: A	162: C
13: B	38: E	63: B	88: C	113: E	138: C	163: D
14: D	39: C	64: D	89: B	114: C	139: D	164: E
15: E	40: A	65: D	90: A	115: D	140: E	165: B
16: C	41: C	66: B	91: B	116: B	141: B	166: A
17: B	42: D	67: E	92: E	117: A	142: B	
18: C	43: B	68: A	93: D	118: E	143: A	
19: D	44: E	69: C	94: D	119: B	144: C	
20: A	45: A	70: D	95: E	120: E	145: A	
21: D	46: C	71: C	96: D	121: C	146: E	
22: E	47: B	72: C	97: A	122: A	147: B	
23: B	48: C	73: E	98: C	123: A	148: D	
24: E	49: E	74: A	99: C	124: D	149: C	
25: A	50: D	75: E	100: C	125: A	150: E	

Copyright © 2015 Sterling Test Prep

Chapter 3

1: C	26: B	51: A	76: E	101: B	126: D	151: A	176: B
2: B	27: C	52: E	77: D	102: C	127: C	152: D	177: A
3: E	28: E	53: D	78: A	103: E	128: B	153: C	178: A
4: A	29: A	54: E	79: A	104: D	129: E	154: E	179: E
5: E	30: C	55: B	80: B	105: C	130: B	155: C	180: D
6: C	31: B	56: C	81: B	106: B	131: A	156: A	181: C
7: A	32: D	57: B	82: D	107: A	132: D	157: A	182: B
8: C	33: B	58: C	83: A	108: D	133: A	158: C	183: E
9: B	34: E	59: E	84: B	109: D	134: B	159: A	184: C
10: D	35: A	60: D	85: B	110: C	135: A	160: C	185: D
11: C	36: A	61: B	86: C	111: B	136: D	161: D	186: C
12: A	37: B	62: C	87: B	112: A	137: A	162: B	187: B
13: D	38: D	63: A	88: C	113: E	138: E	163: D	188: E
14: E	39: B	64: B	89: A	114: C	139: C	164: B	189: A
15: E	40: A	65: D	90: E	115: D	140: A	165: D	190: B
16: A	41: E	66: E	91: D	116: B	141: E	166: D	191: D
17: B	42: D	67: A	92: C	117: E	142: D	167: E	192: C
18: A	43: C	68: D	93: E	118: D	143: E	168: B	
19: D	44: C	69: A	94: C	119: E	144: C	169: E	
20: E	45: E	70: E	95: D	120: C	145: D	170: B	
21: B	46: A	71: D	96: C	121: B	146: B	171: E	
22: D	47: E	72: A	97: E	122: A	147: C	172: A	
23: C	48: C	73: B	98: A	123: E	148: A	173: B	
24: D	49: D	74: C	99: D	124: A	149: C	174: E	
25: B	50: D	75: C	100: E	125: A	150: B	175: C	

Chapter 4

1: D	26: D	51: D	76: C	101: A	126: E	151: A	176: A	201: C
2: C	27: B	52: B	77: E	102: B	127: C	152: D	177: B	202: C
3: E	28: A	53: E	78: D	103: E	128: D	153: B	178: E	203: A
4: A	29: C	54: C	79: A	104: C	129: C	154: C	179: D	204: E
5: E	30: D	55: D	80: C	105: B	130: E	155: C	180: A	205: D
6: B	31: E	56: A	81: B	106: E	131: B	156: D	181: C	206: C
7: E	32: C	57: E	82: D	107: D	132: D	157: B	182: D	207: D
8: D	33: E	58: C	83: B	108: C	133: A	158: E	183: B	208: A
9: D	34: B	59: A	84: C	109: A	134: E	159: D	184: D	209: C
10: C	35: C	60: C	85: A	110: D	135: B	160: C	185: B	210: E
11: A	36: E	61: C	86: D	111: E	136: A	161: C	186: C	211: B
12: A	37: C	62: B	87: E	112: B	137: D	162: E	187: E	
13: E	38: A	63: D	88: A	113: D	138: A	163: A	188: D	
14: B	39: B	64: A	89: D	114: E	139: B	164: D	189: E	
15: C	40: D	65: E	90: D	115: A	140: E	165: B	190: C	
16: A	41: E	66: D	91: E	116: C	141: A	166: C	191: B	
17: C	42: A	67: B	92: C	117: B	142: A	167: A	192: A	
18: D	43: C	68: B	93: B	118: A	143: D	168: E	193: E	
19: E	44: D	69: D	94: C	119: E	144: B	169: A	194: A	
20: B	45: B	70: E	95: B	120: D	145: D	170: E	195: C	
21: A	46: A	71: C	96: E	121: C	146: E	171: B	196: E	
22: B	47: B	72: A	97: A	122: B	147: C	172: A	197: C	
23: C	48: D	73: B	98: D	123: D	148: B	173: D	198: A	
24: E	49: E	74: E	99: E	124: A	149: C	174: E	199: D	
25: B	50: A	75: A	100: A	125: C	150: C	175: B	200: B	

Copyright © 2015 Sterling Test Prep

Chapter 5

1: E	26: C	51: A	76: C	101: B	126: D	151: C	176: C	201: E	226: D
2: B	27: E	52: C	77: D	102: E	127: C	152: A	177: E	202: D	227: A
3: C	28: D	53: B	78: C	103: A	128: A	153: E	178: C	203: C	228: D
4: D	29: A	54: E	79: E	104: C	129: A	154: B	179: A	204: A	229: B
5: C	30: B	55: D	80: B	105: D	130: E	155: D	180: C	205: D	230: E
6: B	31: E	56: A	81: A	106: E	131: C	156: C	181: A	206: E	231: A
7: C	32: B	57: B	82: C	107: D	132: B	157: E	182: E	207: B	232: C
8: A	33: D	58: D	83: E	108: C	133: D	158: A	183: D	208: A	233: C
9: D	34: C	59: E	84: D	109: E	134: E	159: D	184: A	209: D	234: D
10: B	35: B	60: B	85: C	110: C	135: B	160: B	185: B	210: C	235: B
11: D	36: A	61: C	86: A	111: B	136: E	161: C	186: B	211: E	236: E
12: E	37: A	62: B	87: A	112: B	137: A	162: E	187: E	212: D	237: B
13: A	38: E	63: A	88: B	113: D	138: B	163: C	188: C	213: B	238: A
14: E	39: D	64: B	89: B	114: E	139: D	164: D	189: A	214: E	239: C
15: A	40: B	65: C	90: A	115: B	140: B	165: A	190: D	215: A	240: E
16: C	41: A	66: C	91: D	116: A	141: E	166: E	191: E	216: A	241: B
17: A	42: B	67: E	92: E	117: D	142: C	167: A	192: B	217: C	242: B
18: B	43: D	68: A	93: C	118: C	143: D	168: D	193: D	218: B	243: D
19: C	44: C	69: A	94: B	119: A	144: E	169: A	194: E	219: A	244: E
20: A	45: C	70: D	95: D	120: E	145: A	170: B	195: B	220: C	245: A
21: B	46: C	71: B	96: E	121: C	146: C	171: D	196: D	221: B	246: D
22: B	47: D	72: C	97: D	122: A	147: A	172: B	197: A	222: D	247: B
23: A	48: E	73: E	98: C	123: C	148: D	173: C	198: C	223: E	248: A
24: E	49: B	74: D	99: E	124: B	149: C	174: B	199: B	224: C	249: C
25: D	50: D	75: D	100: A	125: A	150: B	175: E	200: D	225: B	250: C
									251: D
									252: A
									253: E
									254: C

Chapter 6

1: D	26: C	51: C	76: D	101: B	126: A	151: C	176: A	201: C	226: A
2: A	27: B	52: A	77: B	102: D	127: D	152: D	177: C	202: A	227: B
3: C	28: A	53: E	78: E	103: E	128: E	153: E	178: E	203: D	228: D
4: E	29: C	54: D	79: C	104: B	129: B	154: B	179: B	204: B	229: A
5: A	30: B	55: B	80: B	105: A	130: D	155: E	180: C	205: D	230: C
6: B	31: A	56: C	81: A	106: C	131: C	156: B	181: D	206: B	231: E
7: D	32: E	57: E	82: E	107: B	132: B	157: A	182: B	207: C	232: D
8: B	33: A	58: B	83: C	108: E	133: E	158: D	183: D	208: A	233: B
9: E	34: D	59: A	84: B	109: A	134: C	159: A	184: E	209: E	234: D
10: A	35: B	60: D	85: B	110: B	135: E	160: E	185: B	210: A	235: C
11: E	36: A	61: C	86: C	111: C	136: A	161: B	186: C	211: B	236: E
12: C	37: D	62: E	87: A	112: D	137: E	162: D	187: D	212: A	237: B
13: D	38: E	63: A	88: B	113: C	138: A	163: A	188: A	213: E	238: E
14: E	39: C	64: C	89: D	114: E	139: D	164: B	189: A	214: D	239: A
15: C	40: C	65: D	90: C	115: D	140: C	165: E	190: C	215: B	240: C
16: B	41: E	66: E	91: E	116: E	141: B	166: A	191: D	216: C	
17: A	42: E	67: B	92: D	117: A	142: A	167: B	192: E	217: B	
18: C	43: B	68: A	93: A	118: D	143: C	168: C	193: B	218: C	
19: E	44: B	69: D	94: D	119: E	144: D	169: D	194: D	219: E	
20: A	45: D	70: B	95: C	120: B	145: E	170: C	195: E	220: C	
21: D	46: C	71: C	96: A	121: A	146: B	171: A	196: C	221: D	
22: B	47: A	72: D	97: E	122: C	147: D	172: D	197: B	222: E	
23: C	48: E	73: E	98: A	123: C	148: B	173: C	198: A	223: D	
24: B	49: D	74: B	99: E	124: A	149: A	174: E	199: E	224: A	
25: D	50: D	75: A	100: D	125: D	150: C	175: C	200: A	225: A	

Copyright © 2015 Sterling Test Prep

Chapter 7

1: E	26: D	51: E	76: B	101: D	126: D	151: C	176: C	201: B
2: C	27: A	52: C	77: C	102: A	127: B	152: A	177: A	202: E
3: A	28: E	53: D	78: D	103: B	128: C	153: C	178: D	203: A
4: C	29: D	54: C	79: E	104: E	129: A	154: A	179: E	204: D
5: B	30: A	55: A	80: A	105: C	130: E	155: B	180: D	205: A
6: E	31: E	56: C	81: D	106: C	131: B	156: A	181: B	206: C
7: B	32: A	57: B	82: E	107: D	132: A	157: C	182: E	207: A
8: C	33: C	58: C	83: A	108: A	133: D	158: E	183: D	208: B
9: A	34: D	59: E	84: C	109: B	134: E	159: C	184: B	209: C
10: B	35: E	60: D	85: B	110: B	135: B	160: D	185: C	210: E
11: D	36: B	61: B	86: E	111: A	136: D	161: B	186: A	211: D
12: E	37: B	62: B	87: C	112: E	137: C	162: D	187: B	212: E
13: A	38: C	63: B	88: A	113: C	138: E	163: E	188: E	213: B
14: B	39: B	64: E	89: B	114: C	139: D	164: B	189: B	214: A
15: B	40: E	65: A	90: D	115: A	140: A	165: D	190: C	215: D
16: C	41: A	66: A	91: A	116: C	141: d	166: C	191: D	216: B
17: D	42: B	67: B	92: C	117: E	142: C	167: D	192: E	217: E
18: A	43: C	68: B	93: B	118: B	143: B	168: E	193: B	218: A
19: D	44: E	69: C	94: D	119: B	144: A	169: B	194: C	219: D
20: E	45: D	70: E	95: E	120: E	145: C	170: A	195: A	220: E
21: B	46: D	71: A	96: D	121: A	146: B	171: E	196: E	
22: D	47: C	72: D	97: B	122: D	147: D	172: A	197: A	
23: C	48: A	73: C	98: D	123: E	148: E	173: D	198: D	
24: E	49: C	74: C	99: A	124: C	149: A	174: E	199: A	
25: A	50: B	75: E	100: E	125: B	150: C	175: B	200: C	

Chapter 8

1: B	26: A	51: A	76: A	101: A	126: B	151: B	176: D	201: C
2: A	27: D	52: D	77: C	102: A	127: D	152: E	177: A	202: C
3: D	28: E	53: E	78: D	103: C	128: C	153: C	178: C	203: E
4: E	29: C	54: B	79: B	104: D	129: E	154: A	179: C	204: D
5: A	30: A	55: D	80: E	105: C	130: C	155: C	180: B	205: E
6: C	31: B	56: B	81: B	106: B	131: D	156: E	181: D	206: A
7: A	32: E	57: D	82: C	107: C	132: A	157: D	182: E	207: B
8: E	33: C	58: C	83: A	108: E	133: B	158: E	183: A	208: C
9: D	34: B	59: A	84: A	109: B	134: C	159: A	184: D	209: A
10: B	35: E	60: C	85: C	110: C	135: D	160: D	185: A	210: D
11: C	36: E	61: B	86: E	111: B	136: E	161: E	186: E	211: B
12: C	37: D	62: A	87: D	112: D	137: B	162: D	187: B	212: D
13: A	38: C	63: E	88: C	113: E	138: E	163: C	188: E	213: E
14: B	39: D	64: D	89: B	114: D	139: A	164: E	189: B	214: D
15: E	40: E	65: A	90: E	115: C	140: C	165: C	190: A	215: A
16: B	41: B	66: E	91: B	116: D	141: A	166: D	191: E	216: C
17: A	42: A	67: D	92: A	117: D	142: D	167: B	192: C	217: B
18: C	43: C	68: B	93: D	118: E	143: B	168: A	193: B	218: E
19: D	44: D	69: C	94: D	119: B	144: C	169: A	194: C	219: D
20: E	45: B	70: B	95: B	120: A	145: B	170: D	195: C	220: B
21: B	46: B	71: D	96: E	121: E	146: E	171: C	196: D	221: A
22: A	47: D	72: A	97: C	122: A	147: D	172: B	197: B	222: C
23: C	48: A	73: C	98: A	123: E	148: E	173: B	198: D	223: B
24: E	49: E	74: B	99: C	124: B	149: B	174: E	199: A	224: A
25: D	50: A	75: E	100: D	125: A	150: C	175: C	200: B	

To access the online SAT tests at a special pricing visit:
http://SAT.Sterling-Prep.com/bookowner.htm

   Copyright © 2015 Sterling Test Prep